Institutional and Industrial Safety Engineering Practices

Scrivener Publishing
100 Cummings Center, Suite 541J
Beverly, MA 01915-6106

Publishers at Scrivener
Martin Scrivener (martin@scrivenerpublishing.com)
Phillip Carmical (pcarmical@scrivenerpublishing.com)

Institutional and Industrial Safety Engineering Practices

Abhishek Barua

Kanchan Kumari

Sumit Saha

Siddharth Jeet

and

Swastik Pradhan

Scrivener Publishing

WILEY

This edition first published 2025 by John Wiley & Sons, Inc., 111 River Street, Hoboken, NJ 07030, USA and Scrivener Publishing LLC, 100 Cummings Center, Suite 541J, Beverly, MA 01915, USA

For more information about Scrivener publications please visit www.scrivenerpublishing.com.

Wiley Global Headquarters
111 River Street, Hoboken, NJ 07030, USA

For details of our global editorial offices, customer services, and more information about Wiley products visit us at www.wiley.com.

Library of Congress Cataloging-in-Publication Data

ISBN 978-1-394-31441-6

Front cover images supplied by Adobe Firefly
Cover design by Russell Richardson

Set in size of 11pt and Minion Pro by Manila Typesetting Company, Makati, Philippines

Printed in the USA

10 9 8 7 6 5 4 3 2 1

Contents

Preface

Safety is one of the key dimensions of monitoring and maintaining the health and well-being of individuals working in a workspace, laboratory, institutional campus, or any industry. Safety by design, or prevention through design, is at the core of keeping systems safe. The objective of this book is to impart knowledge of various aspects and facets of safety, with a focus on tools, techniques, and procedures. This book is divided into two sections: Section 1: Safety and Behavior at Workspaces and Laboratories on Any Institutional Campus; and Section 2: Industrial Safety Engineering Practices.

The institutional campus includes various types of laboratories and workshops that deal with hazardous substances and, at times, hinder the safety of an individual or any assets. As a result, on the premises, certain safety and accident prevention guidelines should be followed and practiced. Section 1 of the book will detail the best safety practices and accident prevention guidelines that should be followed in any laboratory or workshop. It will also detail how to handle any chemicals, materials, apparatus, machinery, or tools, as well as how to dispose of materials, chemicals, or biological waste. It will also address global fire safety practices. The book will also teach readers how to report and react appropriately to accidents, in addition to providing first aid.

Section 2 of the book will provide the concepts of engineering system safety, dimensions of engineering system safety, safety design, and integrating safety with other operational objectives such as quality and dependability. It will also provide some case studies on the practices adapted by industries. Modern industrial safety practices are very different from the literature and information available previously. This book will provide methods that are being used in industries. Safety analytics is used for a variety of purposes, including analyzing safety performance and identifying specific areas for improvement by mapping safety factors based on safety data and incident reports, predicting occupational incidents, and evaluating occupational hazards that must be discussed. Accident investigation and safety will also be discussed, which entails conducting a scientific and academic analysis of the facts that occurred during an accident and conducting an investigation to determine the root cause of the accident in order to make recommendations or take corrective action to prevent the occurrence of the same or a similar event in the future. The rules of occupational safety and health management, which are responsible for preserving worker health and safety, must also be considered.

Authors
Abhishek Barua
Kanchan Kumari
Sumit Saha
Siddharth Jeet
Swastik Pradhan

Section 1
INSTITUTIONAL SAFETY

1

Introduction to Safety: Philosophy and Terminology

Abstract

This chapter deals with the basics of safety and its necessity in an organization. The philosophy of safety, which deals with common causes and their remedial measures, was also discussed. Lastly, some common terminologies associated with safety, i.e., accident, danger, hazards, disaster, emergency, error, and risk, were also stated along with their causes and effects.

Keywords: Safety, philosophy of safety, safety terminology, accident, danger, hazard, disaster, risk

1.1 Background

We live in an unsafe environment. Natural dangers include fire, flood, storm, hunger, and illness. Technological and economic development has made us less vulnerable to these threats. However, technology also introduces new risks, such as workplace mishaps, vehicle accidents, toxic substances, and radiation. Future technologies, such as genetic modification or artificial intelligence, may pose dire dangers to the human species. These dangers are the most compelling case against a naïve or careless approach to development. To completely comprehend development, we must first comprehend danger and safety.

1.2 Introduction

A safe scenario is one in which hazards and conditions are controlled to safeguard people's and society's health and well-being. Individuals, communities, governments, and others must create and maintain the following conditions to achieve an optimal level of safety: a climate of social cohesion

Abhishek Barua, Kanchan Kumari, Sumit Saha, Siddharth Jeet and Swastik Pradhan. Institutional and Industrial Safety Engineering Practices, (3–12) © 2025 Scrivener Publishing LLC

and peace; equity in protecting human rights and freedoms; prevention and control of injuries and other consequences or harm caused by accidents; respect for values; and the physical, material, and social environment. Environmental initiatives (physical, technical, social, economic, political, and organizational) as well as behavioral initiatives can serve to ensure these circumstances.

Safety is a value, and improvements in safety are seen as progress. We have made progress itself safer by requiring higher levels of testing and analysis before new technologies are put on the market. For example, a century ago, little to no testing was performed on new drugs, but, today, they go through extensive, multi-stage trials. In both nature and technology, we must actively pursue and build for safety. It necessitates procedures, norms, and protocols, as well as education and training, feedback loops, leadership, and statistical thinking protocols. Pressure valves, seat belts, and smoke alarms can all assist, but, ultimately, safety needs human answers.

Safety has typically been reactive, but, as we move forward, we must foresee risks in advance. This is because, as the baseline level of risk diminishes, it makes reason to reduce our tolerance for risks of all types, and the more advanced our technology becomes, the more possible harm we can cause. Being vigilant about safety entails recognizing risks through theory rather than experience, and this has intrinsic epistemic constraints. Even when risks are anticipated, individuals do not always take them seriously.

Broad skills help protect against broad categories of risk, including those we cannot predict. Science helps us comprehend risk and what we can do to reduce it; technology provides us with instruments; and riches and infrastructure provide a buffer against disruptions. The route to safety does not involve prohibiting wide areas of R&D or slowing development in general. The route to safety varies greatly depending on the topic. It requires the most accurate threat models that we can create, as well as particular tools, methods, procedures, and standards to counter them.

1.3 Philosophy of Safety

The philosophy of safety is the search for and coordinate common causes and remedial measures for general as well as specific safety problems. It starts with the essential need or prerequisite of safety, its foundations, causation analysis and consequences, methods of detecting unsafe circumstances and unsafe actions, and the reason and order of mishap occurrence. It then investigates physical, physiological, psychological, and other factors affecting and strengthening safety; the costs and types of accidents; the types of safety management; and methods and means of providing and maintaining safe working conditions, as well as human actions affecting people's health and safety and environmental safety. It also discusses the genesis, growth, and amendment of safety legislation, as well as its innovation, as well as all corporate and environmental safety requirements. The field of safety philosophy is limitless, its scope is broad, and it encompasses the entire subject of safety, from its origins to the most recent advances in causality and behavioral analysis to modern concepts of design, testing, reliability, hazard control technology, risk analysis, assessment, and audit, emergency planning, public awareness and involvement programs, and all future developments [1, 2].

The goals of philosophy of safety theory are to consider what is secure, why it is needed, where it is needed, what kinds and applications exist, and so on. The philosophy of safety also consists of protect and service humanity by researching, proposing, and implementing secure behaviors, working circumstances, and environments for everyone's safety, health, and well-being. This demonstrates the utmost significance of the topic of safety, as it has begun to form the age-old need for it eternally. This is the ultimate importance of safety guideline.

1.4 Safety Terminology

Many phrases and terminology in safety science, like in other areas of science, are now well defined. From a legal standpoint, some terms are specified by laws. The following is a list of some frequently used safety terms. It is critical to comprehend these terms because they clarify many ideas in safety theory, science, and law [3, 4].

1.4.1 Accident

An accident is an unplanned and uncontrollable occurrence in which the action or response of a substance, object, radiation, or person causes or increases the likelihood of physical harm. In the public health field, unintentional injury is the favored word for accidental injury. It is defined as any occurrence that disrupts or conflicts with the orderly progression of the activity and causes or is likely to cause harm, with or without property or environmental damage. An accident is defined as an undesired transfer of energy beyond the threshold boundaries that has the potential to cause human harm, property destruction, or both. A fatal accident, also known as a fatality from accident, is defined as an accident that results in one or more fatalities within 1 year of the date of the accident.

Accidents happen either due to any unsafe working condition or to any unsafe act by any person (Figure 1.1). Unsafe conditions include any faults related to mechanical, physical, chemical, etc., whereas unsafe acts include error, ignorance, or overconfidence on the part of a person. Accidents cost an organization directly or indirectly (Figure 1.2). Direct costs include property damage, medical compensation to the affected person, and legal expenses, whereas indirect costs include time loss, production loss, and loss of image for the organization.

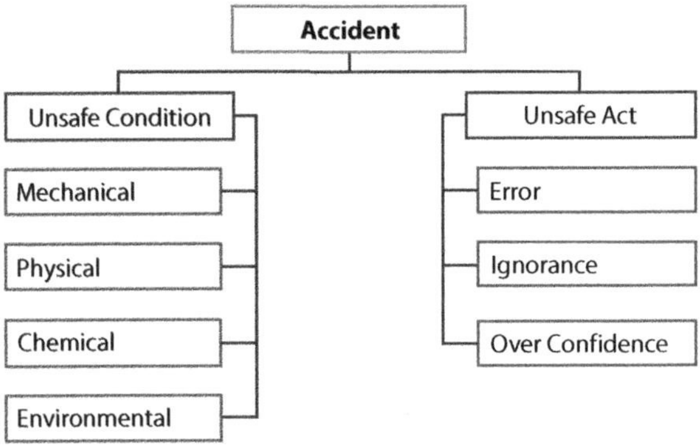

Figure 1.1 Different reasons of accident causation.

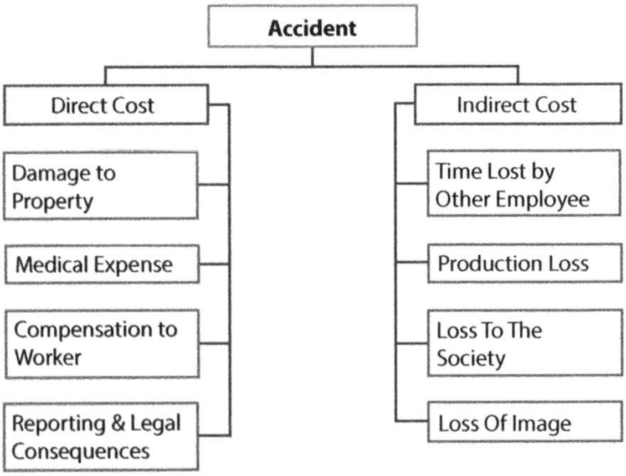

Figure 1.2 Different types of cost due to accident.

1.4.1.1 *Accident-Consequence Analysis*

Accident consequence analysis is an examination of the anticipated consequences of a mishap, regardless of its frequency or likelihood. Typically, this is done after the risk assessment to forecast the outcomes, i.e., the severity of the effects, due to the worst or most plausible accident situations. The risk study of the impacted people, property, and surroundings can be done using accident consequence analysis. Computer software is helpful for performing consequence and risk analyses. In this form of analysis, the determination of the following factors is important:

- The type of material being discharged, for example, gas, liquid with vapor, and liquid.
- The type of discharge, such as instantaneous, sporadic, or constant.
- In the instance of a liquid reservoir, the leak rate, evaporation rate, and discharge amount.
- Calculation of the discharged mass' dispersion: Atmospheric factors are taken into account.
- Damage distances, i.e., the destructive concentration or impact of fire or detonation at various distances in the direction of the breeze or other depicting these distances and effects.
- The severity of the impact, i.e., vulnerability in terms of potential deaths, injuries, structure destruction, or environmental harm. Application of the probit equation.
- Plotting risk markers on a map of the discharge site and surrounding region: Counters should show regions of low, medium, and high risk.

1.4.1.2 *Accident Prevention*

Accident avoidance is both a science and an art. Above all, it symbolizes control, i.e., control over human performance, machine or apparatus performance, and the real surroundings. The term "control" implies both avoidance and rectification of dangerous situations and actions. The first move toward management is prevention. Knowledge of psychology, logic, and management is required to regulate unsafe human actions. Controlling hazardous circumstances necessitates understanding of engineering, health effects, workplace hygiene, ergonomics, and so on. Accident avoidance entails five steps: organization, fact-finding analysis if the facts are discovered, remedy selection, remedy implementation, and review. The sixth monitoring measure should be examined. It includes outcome measurement, appraisal (comparison with legal criteria or standards), feedback, and, if required, further development.

1.4.2 Danger

Danger denotes the degree of exposure to a potentially lethal situation, which can lead to death or serious injury if not avoided. The word "DANGER" should only be used in the most extreme situations (Figure 1.3) and not be used for property damage hazards unless there is also a personal injury risk at these levels. The danger can be reduced by taking appropriate precautions.

Dangerous occurrences can take several forms, including the pressure-induced burst of a steam plant: failure or collapse of lifting equipment, overturning of a crane fire, explosions, and the escape of molten metal, hot spirits, gas, and other substances; and explosion of a pressure vessel, collapse, or subsidence of a structure.

1.4.3 Hazard

A hazard is any item, circumstance, or behavior that can cause injury, sickness, environmental harm, or property damage. Some are readily recognized and rectified, whereas others are unescapable aspects of the work that must be minimized and managed via various control measures.

Figure 1.3 A commonly used "DANGER" sign.

The term "hazards" refers to a wide group. Some hazards are severe and offer an instant threat to one's health and safety. Others, such as many chemicals, vapors, and radioactivity, have effects that take longer to manifest or have accumulated effects.

Physical, chemical, mechanical, ergonomic, biological, and psychosocial are some of the categories of hazards found in a workplace. While each hazard will necessitate a unique set of controls to mitigate the risks associated with it, there is a commonly accepted hierarchy of control measures that applies to all hazards. Hazards at work are not always visible to the naked eye; we must sometimes look closely. There are three types of hazards:

- Visible hazards are those that can be seen, heard, smelled, felt, or otherwise detected by the inspector (open edges, moving machine parts, and vehicles).
- Hidden hazards: Those that cannot be detected without a more thorough investigation (radiation, dust, and ergonomic stressors).
- Developing hazards: Those that worsen over time may go undetected unless measured (erosion and corrosion).

1.4.4 Disaster

Disaster is a catastrophic scenario in which daily routines of humans are abruptly interrupted, leaving them helpless and suffering. As a result, they require security, housing, clothing, medical and other needs of life. Disaster can be natural such as earthquakes, cyclones, floods, and forest fires or man-made such as factory fires, river pollution, industrial accidents, aircraft crashes, explosions, and collisions of vehicles carrying flammable liquids.

1.4.5 Emergency

An emergency is defined as a sudden event that compromises the safety of one or more individuals. To guarantee a fast and effective response, health and safety practitioners must identify possible dangers, minimize risks, and have emergency planning in place. Employees should have a defined framework for dealing with crises while minimizing the danger to human life and health.

The first stage in developing emergency plans is to evaluate risks and vulnerabilities. Practical methods for mitigating risks should be implemented to prevent crises, but staff must be prepared to respond if efforts to avoid an emergency fail. The strategy should include strategies for avoiding harm and death. Structures, machinery, and materials should be protected as much as possible, and the ecosystem and society at large should not be jeopardized. An effective emergency plan will also handle the actions that must be done following the disaster in order for operations to resume. Specific emergency protocols are developed, and employees are taught in their use.

1.4.6 Error

Human error, planning, design error, manufacturing, maintenance, and administration errors are all examples of errors. Human error is described as any action taken by a human that varies from or contradicts prescribed or set behaviors or processes. Human mistake occurs as a result of absence (failure to perform a required function) or commission (performance of a non-required function), failure to recognize dangers, poor reaction, poor timing, the incorrect choice, an abrupt disruption, and so on. It can be either expected or unpredictable.

- Predictable errors arise in comparable circumstances and can be predicted because they have happened several times.
- Random error is unpredictable and one of a kind. For example, a fly or bug may suddenly enter the eye, causing a worker to cast away a tool or lose his equilibrium, resulting in a mistake. However, if flies become a frequent, predictable occurrence, then the error becomes predictable, and corrective steps are needed.

1.4.7 Risk

Risk is the possibility or likelihood that a person will be harmed or suffered an adverse health impact if subjected to a hazard. It is the combination of the likelihood of damage occurring and the degree of harm. The most frequently used meaning of risk in the context of workplace safety and health is that it is the chance or probability of contracting an illness or being injured, while hazard alludes to the agent accountable (i.e., smoking).

The following factors affect the degree or probability of risk:

- The type of exposure or how frequently a person is exposed to a dangerous object or situation
- How the person is exposed
- The intensity of the impact

The possible harm or detrimental health effect of the danger, the number of times people are exposed, and the number of people exposed are used to classify risk categories. For example, exposure to airborne asbestos fibers can cause potentially deadly lung disease, whereas short-term exposure to a display screen may be deemed minimal due to the potential damage or detrimental health effects.

References

1. Fadier, E. and De la Garza, C., Safety design: Towards a new philosophy. *Saf. Sci.*, 44, 1, 55–73, 2006.
2. Hassan, M.H. and Aly Salem, A.M., The Effect of Safety Philosophical Factors on Risk Management. *AIN J.*, 1, 46, 2023.
3. Wilson, L., McCutcheon, D., Buchanan, M., *Industrial safety and risk management*, University of Alberta, Canada, 2003.
4. Gupta, A., *Industrial safety and environment*, Laxmi Publications, India, 2006.

2

Safety and Behavior at Laboratories, Workshop, and Institution

Abstract
This chapter introduces the roles and responsibilities of all the persons associated with any institution for maintaining safety on their premises. How safe lab practices can be and their standard operating procedures have also been discussed. In addition to lab safety guidelines, workshop safety has also been discussed. Different units or shops associated with the workshop where the risk of hazards is high have also been discussed along with their respective safety guidelines.

Keywords: Lab safety, workshop safety, safety guidelines

2.1 Background

Although every workplace and job assignment contains some safety dangers, every institute is vulnerable to hazards that pose a greater risk of extensive property damage, serious injury, or catastrophic environmental harm. Because of the highly specialized nature of the work, many tasks are performed by educated and experienced individuals or under the guidance of specialists. It is important to teach and know about safety, its importance, and how to maintain it at the lower level, i.e., at any institute level.

2.2 Introduction

Institutional safety refers to the practices of safety management that apply to any institute, i.e., school, colleges, research laboratories, and university. Those processes aim to protect students, instructors, researchers, laboratories, machinery, facilities, structures, and the environment from any unwanted hazard or mishaps. All divisions of the institute share responsibility for safety. Various organizations with experience in this area offer

Abhishek Barua, Kanchan Kumari, Sumit Saha, Siddharth Jeet and Swastik Pradhan. Institutional and Industrial Safety Engineering Practices, (13–42) © 2025 Scrivener Publishing LLC

training, technological knowledge, advising help, tools, and services to ensure that all regulatory requirements are met.

2.3 Roles and Responsibilities

A strong culture of safety—a culture in which all people responsibly accept safety as a core value for long-term health and safety—is supported by a shared collaboration between university administration, professors, researchers, laboratory employees, students, and other organizations. The Institutional Safety Responsibility details the precise duties that various institutional administrative leadership, colleges/schools, divisions, primary scientists, lab employees, and others have in relation to laboratory safety. A distinct administrative policy statement explains the duties and obligations of all participants and sets institutional standards for oversight of physical and chemical safety in research and academic settings.

2.3.1 Senior Leadership

The President, Vice President, Provost, and General Counsel, as well as the Vice Provost for Research, make up senior school administration. They are responsible for

- demonstration of all components of safety, i.e., occupational health and safety, environmental health, fire and life safety, and research safety, through providing adequate resources, by public discussion of the importance of safety, and by enacting effective policies to raise knowledge of and ensure compliance with all federal, state, and municipal laws, as well as institution and funding agency policies;
- recognize that hazards occur in a variety of locations, including research labs, campus structures, classroom environments, machine shops, studios, campus grounds, and outdoor study sites;
- be open and honest about who is liable for what, and who is accountable, in all areas of safe work practices, including how study is done in a responsible manner;
- link choices about academic and staff employment, promotions, tenure, and salaries to efforts to encourage safety in the reward and recognition system; and
- take full accountability for safety.

2.3.2 Appointed Institutional Lead

- Engage all parties to develop, carry out, and support a strategy to better the institution's safety culture that is inclusive and collaborative.
- Establish and promote productive working relationships with all parties engaged in enhancing the safety culture.
- Engage in continual evaluation, reporting, development, and dialogue with the campus community.
- Create and gather safety measures that are both intuitive and quantitative.
- Inform top leadership of the culture's obstacles, advancement, and maturation.
- Annual success updates should be made public to the academic community.
- Bring issues up with the top officials.
- Implement recurring evaluations of incidents, procedures, and the safety mindset.

2.3.3 Deans and Unit Heads

Deans, Directors, Chairs, and Assistant/Associate Vice Presidents, who serve in both academic and nonacademic capacities, must

- effectively convey to all employees in academic departments or administrative groups the value of a strong safety culture and the shared responsibility for carrying out work in a safe, healthy, and ecologically friendly way;
- collaborate with academics, employees, and researchers to promote a culture of safety as a shared objective;
- set an excellent example by acting safely yourself; and
- raise issues with top executives.

2.3.4 Faculty/Principal Investigators and Managers

Principal Investigators, professors, and other managers who have direct control over activities are considered managers, both academic and non-academic managers who must

- supervise the safety in research laboratories, campus buildings, teaching environments, machine shops, studios, and field research sites;
- lead by setting a good example of good safety behavior;
- analyze the hazard of any teaching activity, experimental procedure, or work activity that involves any dangerous tool;
- ensure every people receives proper safety training for performing a job safely;
- ensure that all accidents and events involving safety are correctly notified to higher officials; and
- ensure all scrutiny deficiencies are improved.

2.3.5 Students, Visitors, Staff, and Guests

Every employee, postdoctoral researcher, graduate and undergraduate student, and visitor is considered a student, visitor, staff member, or friend who shall

- be aware of possible threats to your and your neighbors safety;
- learn about possible risks connected to your job and your workspace; be aware of where information on these risks is stored for their evaluation and use; and, if the work includes hazardous chemicals, use Safety Data Sheets (SDS) as a guide;
- stop all work, including possibly dangerous laboratory studies, and alert your manager in case of any accident;
- make sure that all incidents or accidents are reported to the higher authority and recorded in the institute's accident and incident reporting database (if available);
- following of all the safety rules, which includes the correct use of personal protective equipment, regulations, engineering controls, and standard operating procedures (SOPs) and safety-related posters, signs, and warning signals;
- participate in emergency drills, and be acquainted with the emergency plan and the emergency assembly location;
- before engaging in any scientific process, or instructional activity, perform a hazard analysis;
- include a hazard analysis as a best practice in your thesis, dissertation, and financing requests for projects pertaining to laboratories;

- any dangerous circumstances should be reported to your supervisor;
- inform colleagues about unsafe conditions and defective tools; take part in the necessary assessment and surveillance programs;
- raise issues with top executives; and
- with your supervisor, fellow researchers, and colleagues, go over the lessons you learned from incidents, mishaps, and near misses.

2.3.6 Department Dealing with Environmental Health and Safety (If Available in the Institute)

- Provide guidance and oversight for raising awareness of safety regulations for protecting the health and safety of all the personnel, including the environment and researchers.
- Collaborate with employees, teachers, and students.
- Assist the institution community in assessing risks and creating policies and other tools.
- Develop best practices and paperwork to communicate institutional standards in collaboration with the institution committees, community, and leadership; offer the institution community access to a central repository of safety resources.
- Observe changes to regulations and advisories and inform the public.
- Raise issues with top executives.
- Collect and submit safety data to leadership, committees, the research community, and the non-research community.
- Provide the study group and the affected employees with training and education.

2.3.7 Laboratory Safety Specialists (If Available)

- For all the research participants, serving as a resource to provide guidance for raising awareness, train, and to ensure observance to al safety regulations.
- Collaborate with the study team and other individuals of the institute.
- Raise issues with top executives.
- Aid the institution community in the assessment of risks, the creation of processes, and the development of other tools.

2.3.8 Selected Laboratory Safety Associates

Laboratory employees and students who have been given this authority by their Principal Investigator are known as Designated Laboratory Safety Contacts. They must

- act as a point of contact between the primary scientist, lab staff, students and other individuals;
- ensure understanding of and adherence to the SOPs, Chemical Hygiene Plan standards, and other laboratory safety and compliance issues by working with the Principal Investigator/Laboratory Director and lab staff;
- with your supervisor, fellow scholars, and colleagues, go over the lessons you learned from mishaps, incidents, and near misses;
- help the principal investigator/lab director with the yearly research safety inspection; assist in organizing and recording laboratory-specific training; and
- elevate your worries to top authorities.

2.4 Safe Lab Practices

Safe laboratory procedures are essential to performing study. Researchers working in scientific labs are exposed to a possibly hazardous atmosphere that has a variety of physical, chemical, biological, and even radioactive dangers. The following are some concerning results from the research that portray a dreadful image of the general mindset toward lab safety [1]:

- According to a study, 30% of scientists say they have seen a lab accident that was serious enough to require medical care.
- Fifteen to 30% of academics claim to have been hurt in a lab mishap or to have experienced a lab-related injury.
- Twenty-five to 38% of lab workers have been hurt or engaged in incidents that were never disclosed to the principal investigator or supervisor.
- When working in the lab, only 40% of poll respondents said they wore personal safety equipment.

- Twenty-five percent of survey respondents reported not having received any training on the particular dangers they face at work.
- Before beginning laboratory work, 27% of academics never conducted any kind of risk evaluation.

2.4.1 Standard Operating Procedures

When performing laboratory work that includes the use of hazardous chemicals, notably extremely hazardous substances, SOPs incorporating safety and health concerns must be created and followed. SOPs are written instructions that specify the actions to be taken during a specific experimental process. They also contain details about possible risks and how they will be avoided. The laboratory staff members who are most informed and engaged in the experimental process should write the SOPs [2].

When creating a SOP, take into account the following information:

- the kind, amount, and makeup of the substance used; keep in mind that the SDS contains crucial details about possible risks that must be taken into account, such as flammability, toxicity, warning characteristics, reactivity, and exposure symptoms;
- the use location, which may include a fume shelter or other confinement equipment; include a space set aside for the purpose of working with especially dangerous chemicals;
- process information;
- the availability of safety tools, such as personal protection gear;
- requirements for waste gathering, storage, and removal; and
- decontamination techniques.

An SDS might not be accessible if the investigation involves the production of a novel chemical. Every attempt should be made to assess possible risks using analogies to related substances or anticipated chemical structure. When the substance's possible toxicity is unknowable, it must be presumed that it is especially dangerous. Laboratory-specific SOPs must cover situations that call for previous permission from the Principal Investigator/Laboratory Supervisor. These conditions are determined by the risks that come with the substance being used, the risks involved in the experimentation process, the worker's degree of experience, and the size of the experiment.

2.4.2 Lab Safety Guidelines

Risky behaviors and a careless attitude toward safety in academics are so normalized that those on the inside are not alarmed or even aware of the poor standards in the field. In order to alter this mindset, the following list of fundamental guidelines for researchers to follow in order to maintain a secure working atmosphere in the lab is provided.

a) Comply with the Guidelines
Before beginning any job, whether it is following a procedure in a book, listening to your instructor or lab supervisor, or following a procedure in class, it is critical to attend, pay attention, and be familiar with all of the steps, from start to finish. Before you begin, clarify any points that are unclear to you or address any concerns you may have, even if they pertain to a later part of the protocol. Before you begin, familiarize yourself with all of the science instruments. If you do not follow it,

- put other lab users and yourself in risk;
- your project is readily ruined;
- place the facility in danger of an accident that might harm staff members as well as machinery; and
- get dismissed or punished.

b) Recognize Where Safety Equipment is Located
Recognizing where the safety equipment is present and how to use it is during critical event is very important. The safety apparatus should be checked in proper interval to ensure its functioning; for example, spraying of water from safety shower and eye wash and working of fire extinguisher (Figure 2.1). Before beginning an investigation, review the lab safety signals and keep an eye out for them.

c) Dress for the Lab
Wearing protective apparel is critical for job safety (Figure 2.2). Gloves, auditory protection, and other items are required, as well as safety glasses and a lab gown. Put on a lab uniform with full arms, closed-toe shoes, and safety glasses before entering the lab. When working in a lab, long hair should be pushed back and out of the way. Remove any jewelry or other metal objects if the task needs extra safety equipment. Wearing protective apparel reduces the risk of contamination while also reducing the risk of skin and eye damage.

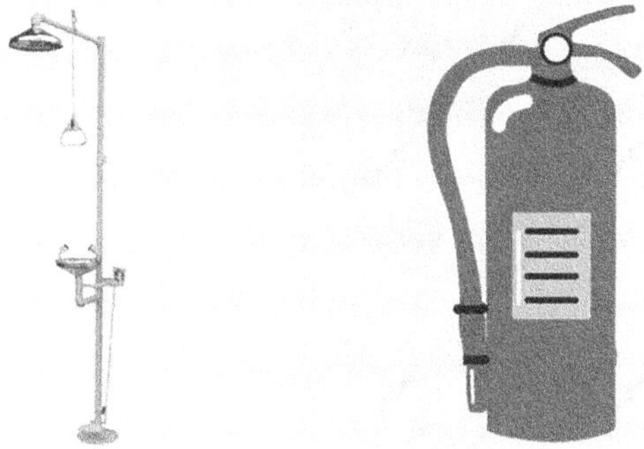

Figure 2.1 Safety shower with eye wash and fire extinguisher.

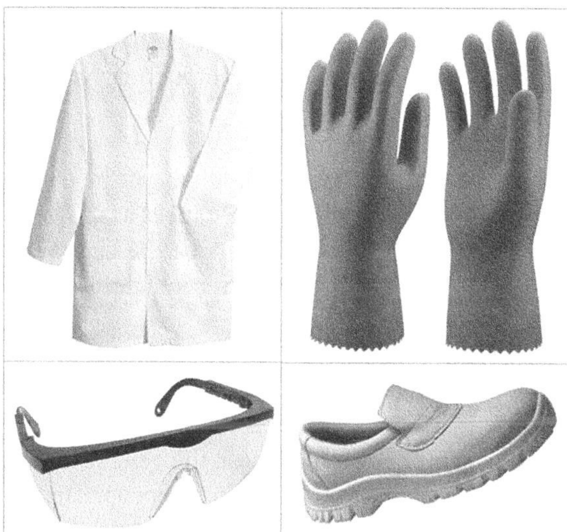

Figure 2.2 Different lab safety attires.

d) Avoid Eating or Drinking in the Lab

Consume no food or beverages in the lab; leave that to the workplace (Figure 2.3). Because the risk of dietary contamination is too high, keep your food and beverages distinct from any studies, medications, or cultures stored in the refrigerator. Handle it with caution if you have been subjected to pathogens or toxins. Drinking in the lab puts your investigation at danger as well. You might get a drink on your lab or study notepad. Distractions like eating and imbibing are allowed in laboratories. You cannot focus on your job if you are consuming.

Figure 2.3 No eating and drinking symbol.

It is feasible to pick and ingest the incorrect liquid if you are used to drinking liquids in the lab. This is particularly true if your glassware is not labeled or if you use scientific glassware as dishes. This is particularly true if you used lab glassware for plates or did not identify your glassware.

e) Avoid Smelling or Tasting Chemicals
You should also avoid sampling or inhaling any of the existing chemicals or biological colonies in the lab. Some substances are toxic or offensive to the senses. Before applying the chemical, name the glassware because this is the easiest way to know what is inside a receptacle.

f) Avoid Behaving Like a Crazy Scientist
Being responsible in the lab is an important safety precaution; avoid acting like a crazy scientist by randomly combining substances to see what happens. As a result, an explosion, fire, or the release of toxic chemicals is possible. Similarly, having joy at work is inappropriate. You may cause harm, annoy people, and even imperil yourself.

g) Proper Disposal of Lab Waste
Before starting an investigation, researchers should have a plan for what they will do after completing the research. When discarding of toxic things or waste, they should use proper disposal receptacles or vessels rather than dumping them down the sink, never return leftover reagents to the container, and dispose of plant detritus in disposal containers. They should also reference laboratory instructions and their teachers or fellow students to determine how sharp items should be disposed of at work.

h) Good Hygiene

- Hands should always be washed after touching any potentially harmful substances, before and after eating, and before exiting the lab.
- Personal belongings should be kept away from lab work, and cosmetics should not be used inside the facility.
- Proper storage containers can help avoid exposure from hazardous substance.
- Chemicals should be kept in non-reactive receptacles, and large quantities of flammable chemicals need to be kept in fire-rated cupboards. Ideal storage for acids and caustics should be in distinct, plastic-lined containers to avoid any vapors from interacting with the metal housing, and waste should be stored in non-reactive receptacles or containers.
- Workspace should be cleaned after conducting experiments.

i) Act Responsibly in the Lab

While experiments are intended for the lab, they should be prepared and studied beforehand. Do not perform arbitrary tests for amusement like the students who drew on the lab table with alcohol and set it on fire to see the lovely blue flame trace the drawing did. When you arrive at the lab, you should be focused and have the correct mindset. A minor diversion can frequently result in permanent harm and, in dire circumstances, even fatalities. When handling possibly dangerous substances or combining compounds, warn other lab participants so they can keep a safe distance. Double-check everything before using it, and make sure to clear up afterward.

j) Carefully Handle Laboratory Apparatus

In addition to chemicals, improper handling of scientific tools can lead to accidents. Be careful when using sharp razor blades, handling heated dishes, and Bunsen stoves. Avoid touching frayed or damaged electrical wires, and handle any shattered crockery with a broom and dustpan. After use, return all tools to their original location.

k) Leave Experiments at the Lab

Do not bring any experimental creatures or chemicals home. You endanger both yourself and them. It is crucial to leave your experiment at the lab for the sake of both your safety and the safety of others. Do not bring it with you home. You might drop something, misplace a sample, or experience a mishap. This might contaminate both the trial and the surrounding area. You might harm someone, start a fire, or forfeit your lab rights in real life. It is recommended to launder the lab-used clothing before using it again.

l) Do Not Experiment on Yourself

Experimenting on yourself can be life threatening to you and your group member. This practice should be avoided.

m) Do Not Work Alone

In a laboratory environment, one should not work alone. In some circumstances, additional senses may spot dangers that you initially do not. Additionally, having nearby experts will speed up assistance in a disaster. This strategy must be changed to accommodate circumstances where it becomes essential to lower the number of scholars in a location.

Prior to accessing the facility and after leaving, alert your supervisor. Make sure everyone is aware of when someone will be nearby by communicating with the nearby laboratories. Try to collaborate in the study with another person. Additionally, two individuals always respond more quickly in case of a mishap. Having someone nearby to get the first aid box or assist with cleaning the glass is helpful, even for a small injury like a laceration from shattered glass.

n) Be Prepared for Lab Accidents

Most labs have a strategy in place for what to do in the case of an accident. Informing a supervisor if and when a mishap happens is a crucial safety guideline. Do not attempt to hide it or fib about it. Accidents do occur, even when measures are taken. In the event of a mishap, remain calm. Anxiety can make things worse. Remain calm and avoid running because you could fall over cables or tip chemical bottles over. Knowing where safety tools like the fire extinguisher, first aid box, emergency phone, and eyewash stations are located is crucial. Wash them off right away if any toxins or particulates end up in your eyes or on your skin. Take lab safety exercises carefully; frequently participating in them will better prepare you for emergencies.

o) Take Part in Safety Drills

Individual researchers are responsible for maintaining a culture of safety and caution at the laboratory, which requires a strong commitment to working attentively, carefully, and safely. They should be aware of the safety tools, the location of the closest fire draw station, extinguisher, spill box, and first aid package, and how to exit and go after a rescue. They should also participate in a yearly safety exercise.

2.4.3 Computer Lab Safety

Every institute has a computer lab, and when using this lab, there are a few guidelines that must be observed. These laboratories house numerous machines and other pieces of equipment, so it is imperative to make sure that nobody does anything that might endanger the machinery there. To ensure that everyone is aware of all the rules that must be followed, the majority of laboratories have their safety warnings posted on a big, noticeable board on the wall. These rules and regulations must be followed, and you can discover some of the most prevalent ones here. With a few small exceptions, these rules and the safety regulations for chemistry labs are comparable in character.

It goes without saying that the lab apparatus in a computer lab is very expensive, and it is your duty to make sure that it is kept secure. The expense of fixing or replacing this equipment will be extremely expensive if it is damaged in any way. Therefore, it is crucial that secondary school students adhere to these lab safety regulations. Knowing this knowledge will enable you to act appropriately in any computer class. Everyone can enjoy themselves in the facility with the proper safeguards in place, and the technology will be protected from harm.

2.4.4 Safe Lab Procedures

The protection in the lab is covered in this part. Safety precautions help shield people from mishaps and harm. They also aid in shielding devices from harm. Some of these rules are intended to safeguard the ecosystem from contamination brought on by inappropriate waste disposal. The following is a sample summary of fundamental safety measures to follow:

- Secure any loose garments and take off your watch and jewelry.
- After using the apparatus, turn off the power and disconnect it.
- Do not touch any hot or high voltage printer areas.
- Understand where to locate and how to use a fire extinguisher.
- Do not bring food or drinks into the office.
- Maintain a clean and clutter-free work environment.
- Electric wires should be kept tangle free and inside a duct.

2.4.5 Electrical Safety

To avoid electrical fires, accidents, and deaths in the house and workplace, follow safety procedures. High voltage is present in CRT displays and power sources. If you are working on power sources or CRT displays, then avoid wearing the antistatic wrist strap. During use, some printer components get heated, and other components could have significant voltage. Do not touch any hot or high voltage printer areas. After using the apparatus, turn off the power and disconnect it. Electric wires should be kept tangle free and inside a duct.

2.4.6 Fire Safety

To safeguard people, buildings, and machinery, adhere to fire safety regulations. Before starting a fix, switch the computer off and unplug it to prevent an electrical shock and computer harm. Fire can cost a lot of money and spread quickly. A minor fire can be kept from spreading out of control by using a fire extinguisher correctly. Be mindful of the potential of an unintentional fire when dealing with computer components and be prepared to respond accordingly. Watch out for smells coming from electronics and laptops. Electronic parts release a burning smell when they overload or brief out. Follow these safety precautions in the event of a fire:

- Never attempt to put out an uncontrolled or uncontained fire.
- Before starting any project, have a prepared fire escape path ready.
- Quickly leave the premises.
- For assistance, call the emergency agencies.
- Before you need to use a fire extinguisher, locate it and study the directions on it. Know the kinds of fire extinguishers that are used in your nation or area. various kinds of flames require various chemicals, which are found in each variety of fire extinguishers.

2.5 Workshop Safety

Every engineering institute or research lab contains a workshop. Also, there are different departments inside those institutes that contain dedicated workshops for them. Departmental workshops are crucial to both study

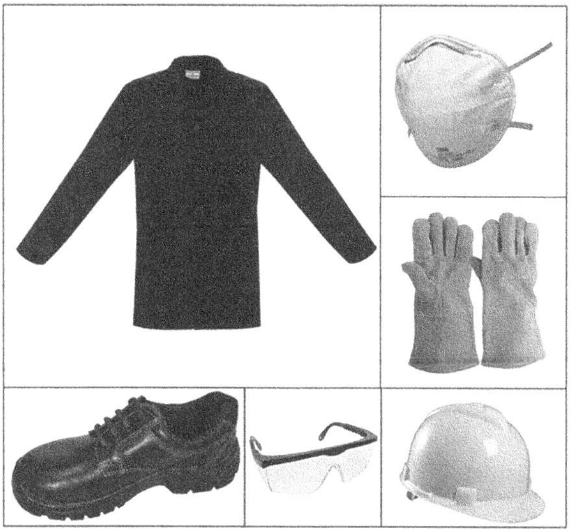

Figure 2.4 Different workshop safety attire.

and instruction. However, using workshop tools improperly can result in severe injuries. The work of students and employees (other than properly trained workshop staff) who are allowed to work in such workshops must adhere to the following protocols in order to reduce the probability of accidents. Outside of an authorized workshop, operating hazardous workshop machinery would not typically be allowed and everyone should be equipped with safety attire (Figure 2.4) [3–5].

The chair or leader of the division must designate a workshop supervisor in divisions that run workshops in order to ensure the workshop is run safely. Only a workshop superintendent may oversee potentially hazardous work that is done in staff workshops. Only duties assigned to staff and pupils in such workshops that have been approved by their supervisor may be completed. The responsibilities of workshop supervisor are as follows:

- Keeping an eye on all work being done in the workshop and making sure that nobody who is not qualified is using hazardous machinery or performing any risky procedures there.
- Maintaining all workshop equipment in a safe working condition and providing workshop safety training to all people.
- Ensuring that proper firefighting, first aid, and other essential emergency equipment is accessible and kept in good working order.

- Ensuring the use of the required personal protective equipment by all the people inside the workshop.
- Regular inspection of the workshop is necessary for identifying the hazards and reporting them to the higher authorities.
- Punctually reporting every accident or incident to the head of the division and the higher authorities of the institution.

2.5.1 General Workshop Safety Rules

- No one else is allowed to labor in this store unless a Workshop Supervisor has given them permission. Power tool operation and other possibly hazardous workplace tasks may only be carried out with a workshop supervisor's approval.
- When using equipment, safety eyewear must be worn at all times. Additional safety gear must be donned as instructed by a workshop supervisor. Wearing neck ties, loose-fitting apparel, dangling jewelry, baggy sleeves, or gloves are not advised. You cannot operate a power instrument until a workshop supervisor has instructed you in safe operation procedures.
- Before making changes to power tools, disconnect the electricity. Avoid maintaining charged devices.
- Take the necessary precautions to stop the discharge of toxic vapors or dust.
- When your job produces heat or sparks or includes exposed fires or flammable substances, take the proper fire protection measures.
- Avoid taking off machine shields.
- Maintain a neat and orderly store. Store materials, tools, and equipment correctly when not in use. After you have finished your task, clean up.

2.5.2 Workshop Policies

- Use the store in groups only.
- Use during workplace supervision hours, or request specific authorization outside of those hours from the shop supervisor.

- Do not do anything that makes you uneasy.
- For instruction, speak with the workplace supervisor.
- Taking note of the particular safety precautions for each power instrument personal safety equipment is on the trolley or rack by the entrance.
- Shoes with closed toes are necessary.
- When construction is being done in the factory, eye safety is needed.
- All slack attire and lengthy hair should be secured.
- To avoid coming into contact with moving equipment, loose hair must be tied back or concealed.
- Burns, blisters, and wounds can be prevented by wearing cutting or work mitts.
- You can choose to use the provided earphones and dust masks.
- It is advised to wear long trousers.
- The store is free of poisonous odors.
- Utilize the outdoor space.
- Keep your work space tidy.
- Inform the workplace superintendent of any injuries.
- There can never be more than 50 pupils working in the store at once.
- Before getting to work, make sure the store is tidy.
- After concluding experiments, clean up.
- Before exiting the store, put the tools back where they belong.
- Clean up liquid spillage right away.
- Get rid of the debris and leftovers.
- Shop vacs and brooms are available.
- No parts of the shop's instruments or machinery may be taken.
- Make sure the door is secured before closing it.
- To prevent injuries, refer to the workplace safety regulations.
- However, a first-aid kit is accessible in the restroom in case of injuries.
- If debris enters the eyes, then wash them completely for at least 15 min at an eye cleansing stand.
- Seek out expert assistance if the issue continues.
- Keep work areas and pathways secure and uncluttered.

2.5.3 Safety Guidelines for Different Shops

There are several shops dedicated to different workshop practices. Hence, it is important to follow safety protocols associated with them. Some of the shops where safety protocols should be strictly followed are as follows:

2.5.3.1 Machine Shop

- At all times, safety eyewear with side screens must be worn.
- Do not operate equipment while wearing open apparel, slack neckwear, or exposed jewelry.
- Work in groups. Always operate in a team.
- Pull back and secure lengthy hair.
- Wear safety shoes.
- Machinists should keep their hands and other body parts away from moving work pieces, machine parts, and blades, and only use hand tools for their intended purpose.
- Report any defective apparatus, equipment, or manual tools to the lab supervisor.
- Never remove or disable any safety guards or mechanisms.
- Never attempt to lubricate, clean, adjust, or repair a machine while it is in use.
- Turn off the machine and lock the power button to the "OFF" position; do not operate any machine unless directed.
- Do not set up or operate equipment unless you have been trained and are familiar with it.
- Never attempt to lubricate, clean, adjust, or repair a machine while it is in use.
- Turn off the machine and lock the power button to the "OFF" position; do not operate any machine unless approved by the current policy and procedures.
- Do not set up or operate equipment unless you have been trained and are familiar with it.
- Only one person should handle a specific piece of equipment and its switches, and it is critical to remove gloves before beginning or operating any equipment.
- Place or manage the material while the machine is turned off.
- Leave tools or work on the machine's surface; remove short, discontinuous types of fragments with a brush; and extricate chips with pliers.

- Never use compressed air to clean tools or clothes, and instead use rags or gloves to clean the work piece or any component of a working machine.
- Always use the correct speed and feeds, and always halt the machine tool when making speed and feed changes that require moving a gear lever.
- A broken implement becomes a hazard and can result in significant physical harm.

2.5.3.2 *Welding Shop*

Welding can be a hazardous profession if safety precautions are not taken. Welders must practice safety by analyzing their surroundings, using tools and equipment as meant, adhering to all maker recommendations, and ensuring that others are not in danger. Before beginning any job, welders must first ensure that they understand how to complete the work safely. Some of the most prevalent hazards related to welding are electric shock, exposure to fumes and gases, fire and explosions, and physical injuries. To avoid these hazards, the following are some safety measures:

a) Avoid Exposure to Fumes and Gases
Adequate ventilation should be provided in the work area to remove fumes and gases from the welding area. When necessary, welders should wear a respirator and express their concerns to a supervisor if they feel their breathing is inhibited.

b) Precautions Against Electrocution
Electric shock is an essential subject for welders to learn about because it can be dangerous. Electrode holder should be checked for any damage before beginning a weld, wearing appropriate gloves, never contact the metal sections of the electrode holder bare hand or body, damp clothing to prevent electrocution. To avoid electrical shock, equipment should be adequately protected. Before welding begins, examine the electrodes as well.

c) Equipment Inspection
Check the equipment before using it, that it is operational and grounded. Even the most experienced welders should inspect their equipment for common wear and tear, such as frayed wires or leaking hoses, as this can increase the likelihood of an accident. Do not assume that a piece of equipment that was working perfectly the day before is still in the same condition. Before using it again, always perform a thorough inspection.

d) Avoid Clutter

A cluttered workplace can cause welding flames and blasts, so keep it free of flammable materials, tools, and equipment. Stay organized and maintain everything in its proper place as a general guideline.

e) Know Your Environment

To guarantee efficiency and safety, take an assessment of your surroundings before beginning a weld. It is critical to know where fire alerts, fire extinguishers, emergency exits, sand buckets, and other tools are located. Welding should be done in a clean, dry area that is free of debris and combustible materials.

f) Dress for the Job

Welders must wear protective clothing because uncovered skin is susceptible to infrared and ultraviolet radiation (Figure 2.5). Additionally, trouser cuffs, pockets, and loose clothing can capture sparks, so they must be fastened. Welders must dress in flame-resistant apparel and use appropriate personal protective equipment such as the following:

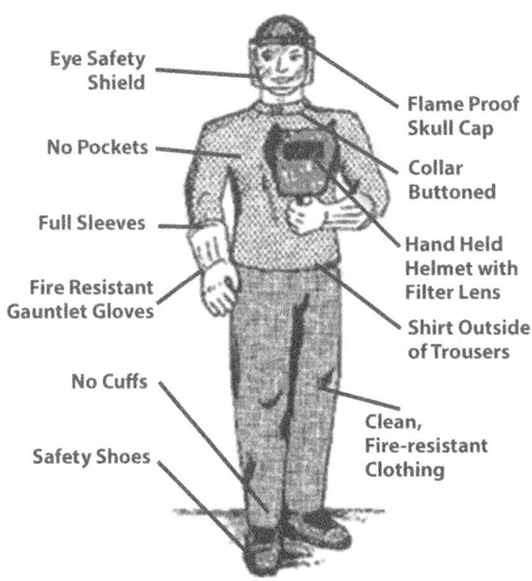

Figure 2.5 Different welding safety attires.

- Ear protection: If noise levels average more than 85 decibels (dB) for eight consecutive hours, you must wear hearing protection at all times.
- Eye and facial protection: This includes safety eyewear, facemasks, and helmets, based on the undertaking.
- Heat and radiation protection: Welders need to wear flame-resistant clothing, gloves to protect their hands and lower limbs, and welding hoods and shields to protect themselves from heat and radiation.
- Fume protection: Welders can be protected from hazardous gases by using fume extraction devices and respirators.
- Electrical shock protection: Welders must wear insulated garments in addition to protect themselves from electrocution.
- Foot protection: The greatest foot protection is provided by leather shoes that are spark and heat proof and have covering above the ankle. Pant legs should be worn over sneakers.

g) Precautions for Fire and Health Safety
Welding is a fire danger, particularly when flammable materials are present. A fire extinguisher and a fire alert must be located near the welding location. Check the capacity on fire extinguishers to ensure that it is full. Standard protocols should be in place to provide immediate assistance to those who find themselves in dangerous circumstances. Health should be prioritized, which includes keeping a completely stocked first aid box handy to treat any injuries as soon as possible.

h) Regular Gas Leak Inspection and Proper Storage of Flammable Products
Gas cylinders should be managed correctly. They should be kept erect, and the regulators should not be shifted between cylinders. Flammable substances and fumes must be kept away from the welding region. They should be kept in specialized storage space.

2.5.3.3 Sheet Metal Shop

Sheet metal is a processed substance that is commonly used in a variety of metalworking uses. Its versatility stems from its malleable and easily handled nature, which enables it to be easily shaped into a variety of forms. Several precautions must be made when working with this material to prevent accidents and injuries. Many safety and health concerns must be addressed when working with sheet metal. These are the following:

a) Use The Proper Equipment

Handling any metal without the appropriate tools can result in harm. Metals transmit energy and heat well, and unfinished edges can be painful to the contact. It can burn and cut the flesh if handled in uncovered palms. As a result, all metalworkers are recommended to use personal safety equipment to safeguard themselves from potential injuries. Although the suggested personal safety equipment will vary based on what the sheet metal will be exposed to during your production process, there are a few standbys. When dealing with metal, employees should wear gloves, safety eyewear, and hard hats at the very least. Steel-toed shoes, long-sleeved shirts, and full-length trousers are also suggested. Workers will be subjected to flying metal debris during some procedures. This debris is made up of chipped-off metal bits that can harm a person's eyes if they come into touch with it. As a result, safety eyewear are also required for these procedures. Depending on how much debris is generated during the process, some locations will also require a complete facial mask to safeguard employees. Metal grinding and welding are two such procedures.

b) Tools Testing

Take the time to try and examine your power tools and machines before using them. It is recommended that all sheet metal processing tools be examined on a frequent basis to guarantee correct operation. This allows people to detect faulty equipment before anyone is injured.

Metal manufacturing and fabrication tools require extensive upkeep because they are exposed to high levels of tension each time they are used. Maintain a routine maintenance plan to ensure that all of your instruments are in good working order. Knowing which instruments require repair or replacement improves working efficiency and safety.

c) Right Technique for Lifting

When dealing with sheet metal, lifting and transporting heavy components is a typical job, but improper lifting methods can result in back and neck injuries. Check to see if the weight is substantial enough to require help before trying a manual lift. When moving things, whether alone or in a group, it is critical to raise with your knees rather than your back and to maintain your spine straight while carrying. Moving as part of a squad necessitates clear dialogue and counting down before moving or lowering sheet metal. Also, keep the others informed if there are any shifts in orientation or possible dangers on the way while transporting the material. Finally, when raising as a group, grasp the material on the edges and have everyone move parallel to it.

d) Clean Up After Handling
The existence of metal particles is a lesser-known danger of dealing with sheet metal. Metal particulates can infiltrate the body through inhalation or absorption, as well as skin penetration in some instances. The risk of this metal dust is the impact it has on our systems. These microscopic pieces can build up in our airways, increasing the chance of respiratory illness and other illnesses. Metal particulates can be lethal in large concentrations. Thus, practicing excellent hygiene before, during, and after a job is critical to remaining secure. Hand hygiene should always be practiced at work during breaks and when exiting the manufacturing floor. Additionally, after use, your tools and clothing should be washed or sanitized.

e) Stay Organized
Keeping the worksite tidy and organized is critical to the safety of all people. Clutter removal has more advantages than maintaining the area clean and organized. Leaving materials and detritus lying around produces tripping dangers. Furthermore, unorganized materials make the operations' process less effective because things are not where they should be. Keep workspaces and pathways free of debris and excess materials. Spills are another item to keep an eye out for. So, if you see a leak, make sure to clear it up right away. Slipping on liquid causes someone to lose their equilibrium, which is particularly hazardous if they are carrying heavy materials. Maintain the practice of storing unused sheet metal between jobs to keep the workspace organized. This keeps the material from accumulating near desks or in pathways. Keeping sheet metal in good condition also contributes to more effective processes. However, when keeping the metal, exercise extreme caution. When arranging sheet metal, make sure the collection is not too large or uneven, or it will topple over.

2.5.3.4 Fitting Shop

Fitting and bench work are essential in the engineering workshop. The following safety precautions should be taken when using hand tools and equipment.

- Keep proper discipline in the shop.
- Do not play with the tools because they are sharp.
- All files should be fitted with a suitable handle at the end of the file.

- When using hacksaw, check its teeth should be sharp and forward position. When using the chisel, keep its direction away from other workers.
- When cutting files forward, only use force in the forward stroke and not the backward stroke, and vice versa.
- When using a hacksaw, make sure the blade is properly inserted into the pins.
- Clean up any spilled oil, grease, or other liquids on the flour; otherwise, they could cause an accident.
- Wear an apron and close-toed shoes at all times; never wear loose clothing.
- No necktie, jewelry, rings, or watches are permitted while working in the shops.

2.5.3.5 Carpentry Shop

While working in a carpentry business, there are some basic safety measures to consider. Some of these are explored further below:

- Check that all of the safety guards are in position and firmly secured before beginning any woodworking equipment.
- When using a circular saw, avoid standing in a straight line with the axis of the revolving blade and keep your fingertips away from the blade's reach.
- The wooden parts should not be fed into the sawing devices quicker than the machine's cutting pace.
- The task should be correctly held while working on wood lathes.
- When performing carpenter labor, one should not use defective or damaged carpentry equipment.
- For proper housekeeping, nails and fasteners should be correctly kept in a box.
- To avoid fire in the carpentry shop, adequate safety measures must be made.
- To save time in dealing, no carpentry equipment should be discarded.
- The teacher or staff person in control of the store will ensure that all necessary safety precautions are taken.
- Only those who have been trained and authorized will be allowed to operate any item of powered machinery.
- Shops must be kept neat and tidy.

- The shop's safety regulations will be displayed.
- Horseplay is not permitted.
- Machines must be examined before use.
- Use the appropriate instrument for the task. Never compel an instrument or attachment to perform a task for which it was not intended.
- Wear no dangling jewelry, mittens, jewelry, watches, ties, ID cards, or anything else that could get stuck in a piece of moving machinery. Long hair should be covered with a head covering, such as a hair net.

2.5.3.6 Foundry Shop

The techniques and materials used in any type of metal casting process are extremely dangerous. Before trying any metal casting, educate yourself on appropriate safety measures. We will go over appropriate safety measures. Consider the following:

- Never pour on damp ground. Do not combine moisture and molten metal, even in tiny quantities. In foundries, steam explosions are the leading source of death.
- Never use water to extinguish a metal fire. This could result in a massive detonation.
- To extinguish flames or contain metal leaks, gather a dry heap of sand and a shovel. All locations should have a sand substrate. This will aid in the containment of metal spillage and the protection of surfaces.
- Spilling molten metal on concrete will cause it to erupt. Over pavement, use a heavy sand layer.
- Use only pure metal as input. Some oils and coatings have harmful combustion residues.
- Always work in a well-ventilated environment. Toxic fumes and dusts can be produced by burning and other industrial chemicals, processes, and metals.
- Wear a particulate shield. Sand, separating dusts, and chemical dusts are all potentially toxic or cancer-causing substances. Take care of your airways.
- Wear protective eyewear at all times. Minor mistakes can result in blindness.
- Never use a crucible that has been thrown or has been harmed. It simply is not worth the danger. Consider what

would happen if a white-hot metal crucible crumbled while you were transporting it. Charge crucibles only when they are frigid. It is extremely hazardous to add metal to a heated crucible. If there is even a haze of dampness on the metal, then it can cause the complete contents of the crucible to erupt.

- Spilled molten metal can move a long way. Maintain a well-defined work space.
- Consider what you are doing at all times. Concentrate on the task at hand and the next move. Prior to any operation, have all movements prepared and practiced.
- Prepare by educating yourself, and always be mindful of your own and others' protection. Read, comprehend, and adhere to the general safety guidelines for the factory outlined above.
- Wear protective equipment. Leather shoes, a fireproof apron, foot and limb protection, appropriate gloves, a wire mesh face mask, safety eyewear, a leather factory cap, and a long sleeve cotton blouse are all required.
- During a fill, only the pouring team is allowed in the furnace. During a deluge, no one should be distracted. Do not gaze into the oven or kilns without a wire mesh cover or proper splattering and infrared radiation eye protection. Only one pour will occur at a moment; the big furnace takes precedence.

2.5.3.7 Fluid Mechanics and Machinery Shop

This shop consists of different arrangements of pumps and turbines driven by electricity and fluid power, respectively. Because water is used under very high pressure along with the usage of electricity, a few safety instructions should be followed in the shop.

- All the machines should be properly insulated so that water will not come in contact with electricity and cause any mishaps.
- All the turbines and pumps should be properly inspected and maintained. They should be operated by professionals and in closed chambers to avoid any water spillage.

- All the pipes should be properly checked for any leakage and repaired.
- People should wear rubber boots and waterproof garments to avoid getting drenched while using any apparatus.
- The shop should be cleaned and dried in case of any water spillage to avoid any accidents.

2.5.3.8 Automobile Engineering Shop

- The protection of automobile engineering shops is essential. Workers will be handling large items, flammable materials, and hazardous substances. They must remember the following safety precautions:
- Organize your business so that your tools and equipment are secure and easily available. To keep the shop clutter-free, securely store tools and car components in cupboards.
- Auto maintenance labor can be dangerously close to a fire. Make sure your shop has fire extinguishers that are readily accessible and suitable for each form of fire: gas, oil, and electrical. Fire detectors should be tested on a frequent basis.
- Maintain a clean and organized work environment so that emergency exits are always obvious and simple to access in the event of a fire.
- Spilled oil, transmission fluid, or other substances can place you in danger of tripping. Make careful to clear up spills as soon as possible.
- To avoid breathing residue from dust particles or toxic refuse, floors should be swept and cleansed on a daily basis.
- To prevent contamination from dangerous substances, never consume or drink on the store floor.
- Follow state regulations when dumping of chemical refuse, particularly flammables like fuel or oil.
- To minimize the danger of breathing toxic fumes, make sure your store is properly ventilated.
- To prevent the risk of collapsing and injuring yourself, never operate underneath a vehicle unless it is properly supported and secure.
- When the engine is operating, never put your hands, tools, or other items near it. The moving components could injure you or cause harm to the equipment.

2.5.3.9 *Thermal/Refrigeration Shop*

Refrigeration, thermal, and air conditioning systems are inherently complex. Only those with the required expertise, training, and equipment should operate and maintain these products. Among the protection precautions are the following:

- Improper compressor handling can result in electrocution, fire due to terminal discharge with combustion, or refrigerant leaks. To prevent severe damage or mortality from electrocution or terminal venting with ignition, take the measures listed below.
- Keep an ear out for arcing noises (sizzling, sputtering, or bursting) inside the compressor. Get away as soon as you hear these noises.
- Contact with refrigerant, refrigerant-oil mixes, or other substances can result in a variety of injuries, including burns and frostbite. For example, if refrigerant comes into touch with the epidermis or eyes, it can result in serious frostbite. In addition, if the compressor engine fails, some refrigerant and oil combinations can become acidic and cause chemical burns. Wear proper protective eyewear, gloves, and apparel when working an air conditioning or refrigeration device to prevent harm.
- If refrigerant or refrigerant-oil mixtures come into touch with your skin or eyes, quickly flush the affected region with water and seek medical assistance.
- Due to the continuous flow of power to the system, heat accumulates in the compressor. This heat will ultimately vaporize the refrigerant, causing the system pressure to rapidly climb. If the thermal protector fails to activate the electrical connection for any reason, then system pressure can increase to levels high enough to cause a compressor housing explosion. After the procedure, turn off the electricity.
- Avoid touching the heat exchanger or fins because they may be hot or have sharp edges. It can cause burns or cuts.
- While operating the equipment, proper ventilation should be present.
- Cooling fans should be properly protected with grills so that no one comes into contact with them.

2.5.3.10 Computer-Aided Design/Computer-Aided Manufacturing Shop

This shop consists of some of the precision equipment, like robotic arms, additive manufacturing machines, and computers. Some of the important safety precautions for this shop are the following:

- The shop should be clean, and the ambient temperature should be maintained for proper functioning of the equipment.
- Because the machines are so expensive, only trained professionals should operate them. Operation of these equipments by any amateur can lead to hazards.
- Additive manufacturing machines run hot during operation. As a result, proper ventilation should be maintained, and no one should touch them because it could result in a burn or other accident. Also they should be operated in closed chamber.
- Robotic arms should be operated in an open space because they can hurt anyone or damage any object.
- Electrical cables should be properly maintained and kept in a duct to avoid any accidents.

2.5.3.11 Material Testing and Measurement Shop

This is one of the most important shops in the workshop for testing and measurement of the physical or mechanical properties of any material. Hence, a few important safety precautions should be followed:

- While testing any material in a machine that breaks them (like the Universal Testing Machine), nobody should be near the machine. Any broken piece may hit anyone and injure them. Those machines should be operated by experts from a safe distance and with the machine guards.
- Nobody should be near the machine while performing an impact test. Only the machine operator will operate the machine safely. Any moving part has the potential to kill someone.
- Proper attire, such as safety shoes, gloves, safety goggles, and an apron, should be worn.

References

1. Ménard, A.D. and Trant, J.F., A review and critique of academic lab safety research. *Nat. Chem.*, 12, 1, 17–25, 2020.
2. Helmenstine, A.M., The 10 Most Important Lab Safety Rules, ThoughtCo, 2021, https://www.thoughtco.com/important-lab-safety-rules-608156.
3. Raghuwanshi, B.S., *Course in Workshop Technology*, Dhanpat Rai and Company (P) Limited, India, 2009.
4. Raghuwanshi, B.S., *A course in Workshop Technology Vol. II (Machine Tools)*, Dhanpat Rai & Company Pvt, India, 2009.
5. Chapman, W., *Workshop Technology Part 1*, Routledge, United Kingdom, 2019.

3

Globally Harmonized System (GHS): Classification and Labeling

Abstract

This chapter explains the globally agreed-upon standard, namely, the Globally Harmonized System of Chemical and Material Categorization and Labeling (GHS). Later parts also provide rules for defining potential hazards of chemical goods and hazard communication tools, as well as material for labels with hazard and precautionary statements, signs, and signal words.

Keywords: GHS, product identifier, signal words, hazard statements, pictograms

3.1 Background

Chemicals have a significant role in the modern world. Still, along with the benefits, they represent considerable risks to people and the environment. Chemical management requires appropriate danger classification and communication strategies, but many countries lack the legislative and technological competence to do so. Chemical hazards are categorized and communicated differently in different countries. As an outcome, employees, consumers, and the environment are not treated fairly. The Globally Harmonized System of Classification and Labeling of Chemicals (GHS) was developed to strengthen and standardize worldwide chemical protection.

3.2 Introduction

The Globally Harmonized System of Categorization and Labeling of Chemicals and Materials (GHS) is a United Nations–controlled, worldwide agreed-upon standard that has been developed to take over the many hazardous chemical categorization and labeling schemes that were previously

Abhishek Barua, Kanchan Kumari, Sumit Saha, Siddharth Jeet and Swastik Pradhan. Institutional and Industrial Safety Engineering Practices, (43–58) © 2025 Scrivener Publishing LLC

used around the world. Standardized hazard testing norms, uniform warning representations, and cohesive safety data sheets provide individuals handling risky objects with a wealth of information. Although it took time, the framework was adopted to varying degrees in almost all of the major countries worldwide by 2017. Prior to the development and implementation of the GHS, multiple independent hazard categorization laws were in use in various countries, resulting in a plethora of norms, categorizations, and labels for the same hazard. Developing a standard that is internationally recognized as a replacement for different regional and local systems offers an opportunity to reduce expenses while increasing compliance [1]. The worldwide norm was meant to replace all of the many categories used in all nations, although it is not a treaty obligation. When member countries implement hazard classification and the Hazardous Communication Standard, they can use the GHS as a common infrastructure.

GHS has two important standardized elements:

1. Hazard classification rules for chemical products (substances, materials, or mixes).
2. Content for labels with risks and precautionary phrases, symbols, and signal words are examples of hazard communication tools.

The GHS was created because several nations had disparate procedures for classifying and identifying chemical items. Furthermore, many systems can coexist within an identical country. While previous methods were comparable in many ways, there were enough variances to result in distinct danger ratings and labeling for a similar product. A product may be classified as carcinogenic in one nation but not in another. This predicament has proved costly for governments to establish and implement, costly for businesses that must comply with several systems, and perplexing for workers who must grasp the risks of chemical substances in order to work safely [2].

3.3 Scope of GHS

GHS applies to all hazardous chemical items, such as those used for the following:

- Industrial chemicals
- Pharmaceuticals

- Consumer chemical goods
- Agricultural chemicals
- Insecticides

GHS's target audiences include employees in a variety of industries (for example, warehouses, construction, chemical manufacture, and transportation), emergency responders, and consumers. The GHS regulates all hazardous chemicals in the workplace at all phases of the product life cycle, including manufacturing, storage, transportation, handling, recycling, and disposal. Depending on national legislation, the GHS may also encompass the use of chemicals in consumer items, which are often utilized in various sectors of the workplace. This can include, for example, chemicals used to sanitize workplaces or paint used in the building industry, to mention a few examples [3, 4].

3.4 Hazards Covered by the GHS

The GHS categorization system is a complicated system that incorporates data from testing, research, and practical experience. A substance's dangers are classified into classes of hazards and severity categories. GHS is divided into three key hazard categories: physical, health, and environmental risks. There are several classifications within each of these danger classes. Figure 3.1 depicts the classification criteria for compounds determined for the hazard.

Physical Hazards

- Aerosols and chemicals under pressure
- Corrosive to metals
- Explosives
- Flammable solids, liquids and gases
- Gases under pressure
- Organic peroxides
- Oxidizing solids, liquids and gases
- Pyrophoric liquids and solids
- Self-heating substances and mixtures
- Self-reactive substances and mixtures.
- Substances and mixtures which, in contact with water, emit flammable gases

Health Hazards

- Acute toxicity
- Aspiration hazard.
- Carcinogenicity
- Germ cell mutagenicity
- Respiratory or skin sensitization
- Reproductive toxicity
- Serious eye damage/ irritation
- Skin corrosion/irritation
- Specific target organ toxicity - single exposure or repeated exposure.

Environmental Hazards

- Hazardous to the aquatic environment (acute and chronic).
- Hazardous to the ozone layer.

Figure 3.1 Hazard group classification.

3.5 Globally Harmonized System (GHS) Labels

The fundamental component of GHS compliance is labeling. With an emphasis on chemical label uniformity and intelligibility, it is critical to understand what gets into a GHS label for a particular container. While correct labeling is the cornerstone of GHS compliance, labels also allow businesses to properly manage and maintain inventories. These labels, on the other hand, must contain the necessary information as well as additional aspects and be clearly comprehended. These regulations affect chemical producers, importers, and distributors. More than 65 nations utilize or are in the process of implementing the GHS system, leading to more transparency in chemical safety usage and transportation. The original chemical label is the finest source of information for learning about dangers, emergency information, and storage. A hazardous chemical GHS label must provide six particular pieces of information (Figure 3.2):

a. Product identifier
b. Signal word
c. Hazard statements
d. Precautionary statements
e. Supplementary information
f. Pictograms

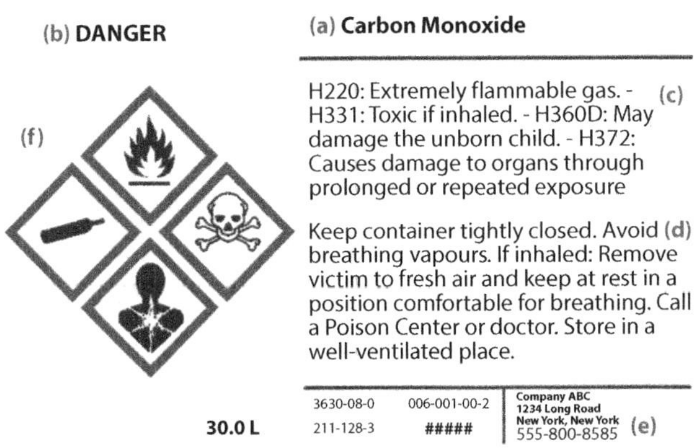

Figure 3.2 Sample GHS label of carbon monoxide: (a) product identifier, (b) signal word, (c) hazard statements, (d) precautionary statements, (e) supplementary information, and (f) pictograms.

3.5.1 Product Identifier

A product identification is used to provide the ID and name of a chemical or product. If other identifiers are required in addition to the product name, then they can be printed on a GHS label to the right of the manufacturer's information. Check that the name and ID on the label and the safety data sheet match.

3.5.2 Signal Words

People can quickly assess the amount of danger associated with a hazardous chemical in a labeled container by using signal words. There are now two authorized GHS signal words in use; however, only one signal word shall be used per label:

- The term "danger" should be used when the hazardous substance being labeled poses an immediate and substantial risk, such as death or long-term health concerns.
- The word "warning" should be used when the dangerous chemical requires lengthy exposure to cause health problems.

3.5.3 Hazard Statements

Hazard statements are GHS-compliant statements that describe the hazards that a substance poses. A hazard statement, for example, should inform the reader whether the materials are hazardous to their health or body. Each danger categorization category has its own set of statements.

- Describes the type and degree of the hazard (if possible): A hazard statement, for example, should inform the reader whether the materials provide a physical threat or a health concern.
- When required, statements can be merged. Some documents may include many statements, while others may have only one.
- Statements are tailored to certain danger categorization groups.
- Cautionary statements: Cautionary statements indicate what persons who use the material should do to limit or avoid exposure to hazardous substances. Precautionary statements are classified into four types: prevention, emergency

Physical Hazards

- **H200 - H208:** Unstable explosives
- **H221 - H228:** Flammable gas/vapour/liquid/solid
- **H230 - H232:** Sensitive to air exposure
- **H240 - H252:** Heating may cause fire
- **H260 - H261:** Unstable in the presence of water
- **H270 - H272:** Oxidizer
- **H280 - H281:** Gas containment issues
- **H290:** Corrosive to metals

Health Hazards

- **H300 - H305:** Harmful if swallowed
- **H310 - H320:** Causes skin and eye issues
- **H330 - H336:** Harmful if inhaled
- **H340 - H341:** Genetic defects
- **H350 - H351:** May cause cancer
- **H360 - H362:** Fertility
- **H370 - H373:** Organ damage

Environmental Hazards

- **H400 - H401:** Toxic to aquatic life
- **H402:** Harmful to aquatic life
- **H410 - H413:** Toxic/harmful to aquatic life with long lasting effects
- **H420:** Ozone destruction
- **H433:** Harmful to terrestrial vertebrates

Figure 3.3 Hazard categories and their codes.

response, storage, and disposal. "Wear protective goggles" is an example of a precautionary remark. A "do not induce vomiting" reaction might be appropriate in an emergency.

The categories and codes (Figure 3.3) that can be used in a danger statement are as follows: Each of the three categories has numerous particular codes that must be chosen based on the qualities of the chemical:

- Chemical dangers that produce a physical hazard, such as a fire, explosion, or other concern, are classified as physical hazards.
- Chemical dangers that create a health concern when swallowed, breathed in, applied to the skin, or in other scenarios are referred to as health hazards.
- Chemical risks that have a detrimental impact on aquatic environments are referred to as environmental risks.

3.5.4 Precautionary Statements

Precautionary statements tell readers what procedures to take in the case of an exposure and how to reduce or prevent dangerous exposures. Figure 3.4 depicts the five various forms of cautious statements that can be used on a GHS label, as well as their "P" code formats.

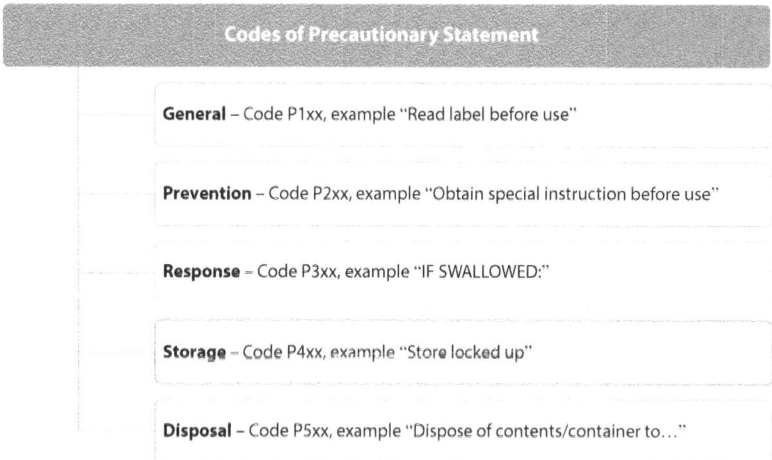

Figure 3.4 Precautionary statements codes.

3.5.5 Supplementary Information

This part is not included in the list of necessary label sections because many items will not require it. Additional information should include the following:

- Additional directions for handling the item that are not addressed by the GHS precautionary statements should be included in the supplemental information.
- Some chemical dangers are not categorized under the GHS criteria, but they must still be stated on labels and safety data sheets.
- This section must specify the proportion of ingredient(s) with uncertain acute toxicity when present in less than 1% concentration.
- Manufacturer details: This includes the name of the manufacturer's firm as well as contact information such as an address and phone number.
- Product name or identifiers: This is the name of the product or chemical. Aside from the manufacturer's information, further identifiers might be noticed.

3.5.6 Pictograms

Pictograms are standardized symbols that are used to swiftly provide material handlers with an image of the types of threats the substance provides. These pictograms are used to indicate hazardous materials and are often classified according to chemical or physical danger, health risk, or environmental risk. A pictogram is made up of a sign and other features that communicate information, such as a border, background pattern, or color. Diamond-shaped danger pictograms should be used for all hazards. Are usually framed by a red square frame that is positioned on point. The picture is usually a black symbol on a white backdrop. This implies that some manufacturers will have to improve their equipment in order to print color labels, or they will have to start utilizing pre-printed labels. Pictograms must have a minimum dimension of 3.9 inches by 3.9 inches for the transportation of hazardous chemicals, with the exception of smaller pictograms on extremely compact containers and gas cylinders. The symbols on transport pictograms are located on the upper side of the label. The pictograms should be attached to a contrasting color backdrop. There are 11 different pictograms available, each reflecting a different danger in one or more categories:

3.5.6.1 Chemical Risk Pictograms

a) Corrosion
The corrosion pictogram (Figure 3.5) recognizes corrosive metals as well as compounds that, due to their corrosive qualities, offer the danger of eye injury or skin corrosion. If inhaled, then it might be dangerous. This pictogram is also used in the area of health risks.

b) Gas Cylinder
The cylinder symbol (Figure 3.6) specifies use for any under-pressure gases. This comprises pressurized gases, liquid gases, and dissolved gases. Because the gas is being kept under pressure in a cylinder, there is an explosion risk. If heated or dropped, then its container may explode.

c) Flame
The flame pictogram distinguishes flammable and self-heating materials, as well as self-reactive and organic peroxides, as well as pyrophoric and flammable gas emitters (Figure 3.7). This comprises not only typical flammable chemicals like gasoline but also liquids and solids that burst into flame when exposed to air and water, as well as mixes that have the ability to self-heat and spontaneously ignite.

Figure 3.5 Pictogram of corrosive material.

Figure 3.6 Pictogram of gas cylinder.

d) Flame Over Circle

This pictogram depicts a fire burning over the letter "O" (Figure 3.8). It is a term used to describe gases or liquids that oxidize or lose electrons. This may appear to be a small issue in comparison to explosives, yet the oxidization process may radically change a material. Oxidizers are substances

Figure 3.7 Pictogram of flammable substance.

Figure 3.8 Pictogram of oxidizing substance.

that help flames burn hotter and for longer periods of time. When it comes into contact with flammable objects, it may react aggressively or explode.

e) Exploding Bomb
This pictogram (Figure 3.9) denotes explosives, such as organic peroxides and extremely unstable materials that can explode even when not exposed

Figure 3.9 Pictogram of explosive.

to air. It also includes self-reactive compounds and mixes, which can react explosively to even tiny physical shocks or even a slight rise in air pressure.

f) Reactive Material
This symbol (Figure 3.10) denotes dangerously reactive materials. A material is considered dangerously reactive if it exhibits three distinct properties

Figure 3.10 Pictogram of reactive material.

or abilities: first, it can vigorously react with water to produce a toxic gas; second, it will react with itself when shocked (bumped or dropped); and third, it can vigorously polymerize itself, decompose, or lose extra water to become a more condensed material. Shock, friction, temperature rise, or heating in a closed container may cause it to explode.

3.5.6.2 *Health Risk Pictograms*

a) Acute Toxic
This exclamation mark (Figure 3.11) pictogram indicates skin and eye irritants, skin sensitizers, compounds that provide a risk of acute toxicity, those that may have narcotic effects, and those that may induce respiratory tract infections. Acutely hazardous compounds produce damage quickly, may be absorbed through a variety of routes (including breathing and skin contact), and can harm the eyes, skin, and internal organs even with minor exposure. This symbol is also used for products that may endanger the ozone layer, although this is optional.

b) Health Hazard
This pictogram denotes a material that, when breathed, poses a health risk, has high carcinogenicity, and/or directly destroys one or more organs (Figure 3.12). This pictogram also represents materials with a risk of reproductive toxicity as well as those with a risk of mutagenicity.

Figure 3.11 Pictogram of acute toxic material.

Figure 3.12 Pictogram of substance that possess health hazard.

c) Severe toxic

The skull and crossbones symbol (Figure 3.13) acts as a warning about drugs that are immediately and severely hazardous. These are drugs that can cause considerable harm with only a single dosage taken orally or through the skin. Substances such as poisons and extremely concentrated

Figure 3.13 Pictogram of severe toxic material.

acids, which have the potential to be lethal poisons and inflict irreversible harm if breathed, eaten, or entered the body through skin contact.

3.5.6.3 *Environmental Risk Pictograms*

a) Environment

The environmental risk pictogram (Figure 3.14) is used to designate items that are potentially hazardous to aquatic life. This symbol, as the dead fish suggests, warns that a material is hazardous to the aquatic environment and should not be permitted to reach lakes, streams, or groundwater. This can include both acute hazards that inflict significant immediate harm to the environment and chronic hazards that can ruin an ecosystem for years or decades.

b) Bio-Hazardous

This pictograph resembles three "C's" connected together with a little circle in the center, all within a circle (Figure 3.15). It is used to mark biological objects that pose a serious health danger, such as virus samples and discarded hypodermic needles. A biological hazard, often known as a biohazard, is a biological contaminant that endangers the health of living creatures, particularly humans. It has the potential to produce a major disease, resulting in sickness or death.

Figure 3.14 Pictogram of substances that possess environmental risk.

Figure 3.15 Pictogram of bio-hazardous substance.

References

1. Nations, U., Globally harmonized system of classification and labelling of chemicals (GHS), United Nations, New York, NY, USA, 2011.
2. Pratt, I.S., Global harmonisation of classification and labelling of hazardous chemicals. *Toxicol. Lett.*, 128, 1-3, 5–15, 2002.
3. Boelhouwer, E., Davis, J., Franco-Watkins, A., Dorris, N., Lungu, C., Comprehension of hazard communication: Effects of pictograms on safety data sheets and labels. *J. Saf. Res.*, 46, 145–155, 2013.
4. Winder, C., Azzi, R., Wagner, D., The development of the globally harmonized system (GHS) of classification and labelling of hazardous chemicals. *J. Hazard. Mater.*, 125, 1-3, 29–44, 2005.

4

Safety Data Sheet (SDS)

Abstract
This chapter discusses the use and importance of the Safety Data Sheet (SDS), which contains crucial occupational safety and health information. It also explains how the SDS is prepared by the manufacturers, where all the vital information about the chemical or material is given so that the person using it can handle it properly. Lastly, all 16 sections of SDS are also discussed in detail.

***Keywords*:** Safety Data Sheet, hazard information, handling

4.1 Background

It is critical for a person to take basic safety precautions, especially while working with various types of materials. There is a substantial danger of harm or disease if the individual is not familiar with the health implications of exposure to a certain item. A Safety Data Sheet (SDS) contains technical specifications, including extensive details regarding a product, and emphasizes important aspects such as an assessment of the hazards associated with storing and handling the item; the possible consequences for health in cases of being exposed; emergency protocols in case that item causes injury or illness; and any measures that handlers must take to mitigate exposure.

4.2 Introduction

An SDS or Material SDS is a standardized document that contains crucial occupational safety and health information. It is mandated by the International Hazard Communication Standard (HCS). According to this standard, chemical manufacturers must communicate the hazard information of their chemicals to those who handle them, and one way to achieve this is by providing a SDS. These sheets are comprehensive and cover

Abhishek Barua, Kanchan Kumari, Sumit Saha, Siddharth Jeet and Swastik Pradhan. Institutional and Industrial Safety Engineering Practices, (59–72) © 2025 Scrivener Publishing LLC

essential details such as chemical properties, probable issues pertaining to health and the environment, appropriate safeguards, and preventative measures for appropriate chemical storage, handling, and transportation.

SDSs are documents created by chemical makers or suppliers that provide information about the material's physical and chemical qualities, potential risks, and how to operate carefully with these substances. They also provide information on the dangers of the materials' consumption, keeping, handling, and emergency measures. In reality, they serve as a single source of information on dangerous compounds. SDSs are primarily intended for use by individuals who work with hazardous chemical substances. SDSs must be revised at least every three years, but if significant new information on the item becomes available, it should be added to the SDS prior to the 3-year deadline.

4.3 Importance of Safety Data Sheet

Safety Data Sheets are essential for assisting anyone in making the workplace safe and protecting the environment. SDS, in particular, offers information to assist you in making a risk assessment. If what is required fails to appear on the SDS, then the supplier should be contacted. SDS is not an assessment in and of itself. It will, however, define the dangers, allowing you to determine the likelihood of those hazards occurring in the workplace.

4.4 Who Produces Safety Data?

For risky substances or products, the supplier is accountable for delivering the correct safety data sheets. The manufacturer, importer, or distributor may be the provider. These entities have the primary responsibility for assessing the hazards of their products and providing relevant safety information to downstream users. SDS should not be received after the product. It might, however, be provided separately, for example, electronically. It is also critical to distribute SDSs with adequate information to ensure that the materials or chemicals may be used safely in the future. The production of SDS involves conducting thorough evaluations of the physical, chemical, and toxicological properties of the substance. This includes determining its potential hazards, such as flammability, reactivity, and health effects. The manufacturer or supplier also considers the intended use of the substance and any specific handling or storage requirements. Based on this assessment, the producer prepares the SDS, which includes standardized sections and information as prescribed by regulations or industry guidelines. It is important to note that, whereas the producer is responsible for creating the SDS, downstream users, such as employers, are responsible for obtaining and ensuring the availability of the SDS for their users and ensuring compliance with applicable regulations. One might be able to create their own SDS using the SDS provided by the vendors. However, it is the user's obligation to ensure that the SDS is correct. Even though the SDS is sufficient when received, they may need to provide more details if they are more aware of how the good is going to be utilized than the original source.

4.5 Need of Safety Data Sheet

Overall, only SDSs for hazardous substances are required. An exemption is allowed for some substances or materials that are not considered harmful. Suppliers must make an SDS accessible upon request if they contain a harmful chemical or substance at a particular concentration or have an acceptable exposure limit. Many vendors currently offer SDSs for unclassified substances. Here are some key reasons why SDSs are needed:

a) Hazard Communication: SDSs play a vital role in hazard communication by providing essential information about the potential hazards of a substance. They contain details about physical, chemical, and toxicological properties, as well as information on health effects, fire hazards, reactivity, and environmental impacts. This information helps users understand the risks associated with the substance and take the necessary precautions to protect themselves and others.
b) Safety and Risk Management: SDSs provide guidance on the safe handling, storage, and transportation of hazardous substances. They outline proper storage conditions, recommended personal protective equipment (PPE), and spill control measures. By following the instructions and precautions mentioned in the SDS, individuals can mitigate risks, prevent accidents, and promote a safe working environment.
c) Emergency Response: SDSs include instructions for responding to accidents, spills, leaks, or exposure incidents. They provide guidance on first-aid measures, firefighting procedures, and proper cleanup and disposal methods.

In the event of an emergency, SDSs serve as a quick reference to ensure a timely and appropriate response, minimizing potential harm and damage.

d) Regulatory Compliance: SDSs are a legal requirement in many countries and are mandated by regulatory bodies such as the Occupational Safety and Health Administration in the United States and the European Chemicals Agency in the European Union. Compliance with these regulations ensures that businesses and organizations are meeting their obligations to provide a safe working situation and protect the worker.

e) Information for Users: SDSs are valuable sources of information for individuals who work with or are exposed to hazardous substances. They help users understand the potential health effects and risks associated with a particular chemical, which aids in making informed decisions about its use, storage, and disposal. Additionally, SDSs often provide information on engineering controls, exposure limits, and personal protective measures, enabling users to take appropriate precautions to minimize risks.

Overall, SDSs are essential tools for promoting safety, health, and environmental protection. They provide comprehensive information about hazardous substances; guide users in their safe handling, storage, and use; and facilitate effective emergency response when necessary.

4.6 Responsibilities Related with the Safety Data Sheet

Various responsibilities are associated with the SDS, involving different stakeholders. Here are the key responsibilities related to SDS:

4.6.1 Manufacturers, Importers, and Distributors

- Preparation: Manufacturers, importers, or distributors of hazardous substances are responsible for preparing accurate and comprehensive SDSs for their products. This involves conducting hazard assessments, gathering relevant information, and ensuring compliance with regulations.

- Content Accuracy: They have to verify that the material in the SDS is correct, is up to date, and represents existing awareness and comprehension of the dangers associated with the substance.
- Language and Format: SDSs should be made available in the official language(s) of the nation in which the chemical is sold or used. They should follow the prescribed format, including standardized sections and headings.
- Distribution: Manufacturers, importers, or distributors must provide SDSs to downstream users, such as employers, upon supply or sale of the hazardous substance. This can be done in physical or electronic form.

4.6.2 Employers and Users

- Access and Availability: Employers who use hazardous substances in the workplace are responsible for obtaining SDSs from suppliers and ensuring their availability to employees. SDSs should be readily accessible in the workplace and easily retrievable.
- Employee Training: Employers must train employees on the hazards associated with the substances they work with and provide information on how to read and interpret SDSs. This training helps employees understand the risks, appropriate handling measures, and emergency response procedures.
- Hazard Communication: Employers are responsible for effectively communicating the information from SDSs to employees, including specific hazards, safe handling practices, necessary PPE, and other control measures.
- Record Keeping: Employers should maintain a record of SDSs received from suppliers and make them available for review by employees, occupational health professionals, and regulatory authorities.

4.6.3 Regulatory Authorities

- Enforcement: Regulatory authorities have the responsibility to enforce compliance with regulations related to SDSs. They may conduct inspections, audits, and take necessary actions to ensure manufacturers, importers, distributors, and employers meet their obligations regarding SDSs.

- Review and Approval: Regulatory authorities may review and approve SDSs for certain hazardous substances or industries to ensure compliance with specific requirements or to address unique hazards.

It should be noted that the precise obligations connected with SDSs might vary based on the nation, legal restrictions, and type of chemicals that are utilized or handled. Compliance with these duties encourages the safe transportation, storage, and use of toxic substances while also protecting both people and the ecosystem.

4.7 Contents of Safety Data Sheets

SDS must contain specific information to ensure comprehensive communication of hazards and safety measures associated with hazardous substances. The content and format of SDSs are typically governed by regulations such as the Globally Harmonized System of Classification and Labeling of Chemicals (GHS). SDS has 16 categories in all. The information under these topics will assist you in ensuring that that item is used responsibly. The SDS, for example, should define the threats that the chemical may present, offer guidelines for the way it ought to be handled, stored, and disposed of, and clarify what is required in the event of an accident, such as first aid and firefighting techniques. Despite the fact that these titles are required, the preparation of SDS provides recommendations on the type of details that should be included under the headings. However, it is the supplier's obligation to assure that the user has adequate information to make an informed decision about how to safeguard workers and the environment at work. While specific requirements may vary slightly depending on the jurisdiction, SDSs generally include the following sections:

4.7.1 Identification of Product and Company

The SDS includes the product identification (often the product name). To find the proper SDS, consistently use the product identification rather than an abbreviated. Make sure the manufacturer's/supplier's names are included on the label. Other identifiers, including a product code or brochure number, may be displayed on the SDS and label.

4.7.2 Identification of Hazards

The hazard identification portion outlines how the substance can be exposed and the detrimental health impacts it can have. Effects seen in test subjects might be mentioned when they are thought to be relevant to humans.

a) Regulatory Status
   This subsection may provide information on the material's regulatory status under the Controlled Products Regulations.
b) Emergency Overview
   This area describes the material's appearance, e.g., smell, physical shape, color, as well as the most serious immediate problems, such as reactivity, fire, and health and environmental dangers.
c) Potential Effects on Health
   i. *Primary Exposure Routes:* Exposure can happen through eye contact, ingestion, skin contact, or inhalation. When developing strategies to reduce exposure, each point of entry must be evaluated. Chemicals taken into the body can have far-reaching effects on bodily systems and organs. For example, phenol can induce deadly nervous system and renal problems if absorbed via the skin.
   ii. *Effects of Acute Product Exposure:* An acute exposure occurs over a brief period of time. The health consequences of acute exposure are generally visible over a time period. One must be aware of the usual effects of a brief exposure in order to avoid being mistakenly exposed. Any symptoms related to the use of a substance should be reported followed by root cause investigation. The causes of the symptoms might vary greatly.
   iii. *Effects of Chronic Product Exposure:* Such exposures can be classified as either prolonged (extremely long) or repetitive (many exposures). Any ailment caused by chronic exposure may manifest slowly or not at all until a long time after the contact has ended. One must be aware that, while there may be no warning symptoms at the time of exposure, a disease presumably connected to the contact may arise later. If these sorts of consequences are feasible for the substance being handled,

then it is critical to limit their exposure by adhering to recognized safe handling standards.

Because a single item will not impact everyone in the same manner, the medical information in the Risk Assessment subsection of the SDS must be regarded as generic. If the material includes carcinogens, skin sensitizers, or respiratory sensitizers, then adopting proper handling practices and processes is critical to minimizing your exposure.

d) Possible Environmental Consequences

This section explains the material's possible impacts if it is released into the environment, such as if it would wildlife and aquatic life or if it would accumulate in the ecosystem.

4.7.3 Information on Composition

This section lists potentially dangerous chemical substances, byproducts, and impurities in the product, as well as the estimated amount (%) of each. Chemical Abstracts Service codes for the substances are also frequently supplied. This part may also specify whether any or all of the parts are protected trade secrets.

4.7.4 Measures for First Aid

This section outlines what to do right away if you are accidentally exposed to the material. Before working with the substance, it is necessary to be familiar with first-aid procedures. During an emergency, there is not enough time to locate and read the SDS. First-aid techniques should be reviewed on a regular basis, especially by people who have received first-aid training. Everyone should be aware of the location of first-aid facilities and equipment, such as safety showers, eyewash fountains, and first-aid kits. The SDS should be sent to the emergency facility if it is readily obtainable when medical care is required for the victim. If the SDS is not accessible, then submit the substance's label or, if small enough, a labeled container of the material. Emergency medical personnel must be aware of the nature of the substance as well as the first-aid actions that have been advised and implemented. The SDS may contain additional instructions that the emergency doctor may find valuable.

4.7.5 Measures for Fire Fighting

This section discusses any potential fire dangers related to the material as well as firefighting measures. This section's information can be used to decide where a particular substance ought to be kept to avoid mishaps.

4.7.6 Measures for Unintentional Release

This section contains instructions for reacting to an unintentional discharge or clearing up a leak. The material is primarily intended for use by rescue workers and environmental specialists.

4.7.7 Storage and Handling

This section includes instructions for appropriate substance handling as well as any gadgets that may be necessary. Much of the material in this part is aimed at health and safety specialists at work as well as those in charge of constructing safe handling and storage facilities.

4.7.8 Personal Protection and Exposure Controls

This section contains data that may be used to create protocols and practices for dealing safely with the substance.

a) Course of Action for Exposure: Exposure guidelines are provided for every element. These are often applied exposure constraints, such as baseline limit values. When doing air sampling, health and safety experts use exposure limits as benchmarks.
b) System for Engineering Controls: These systems mitigate possible dangers by separating or eliminating them from the work area. The next best choice is engineering control systems, which are recommended over other control methods such as the application of personal safety gear. When dealing with the material, one must ensure that the engineering control mechanisms advised for their job are correctly examined and maintained and that they are operational. If the procedure or materials change, then the controls may need to alter as well.

c) Protection for Eyes: Depending on the task and the substance being handled, different types of eye protection (e.g., safety glasses or a face shield) may be required.
d) Protection for Body: Gloves, full-body suits, aprons, mask, and boots are examples of skin protection. The SDS ought to inform us what kinds of rubber or additional supplies are ideal for protecting against the product we are employing. There is no one substance that functions as an obstruction to all chemicals. It is also crucial to think about the temperature and the necessity for fabrics that would not rip or tear easily. There is general advice on the necessity of and choice of personal safety gear.

4.7.9 Chemical and Physical Properties

The specification of the item on the SDS should match the substance it contains. If it does not exist, then the SDS is improper. Instead, the substance might be old or have deteriorated while being shipped or stored. In either situation, the details on the SDS could not be applicable, and they should seek further guidance. The remainder of this section's information is utilized to assist in evaluating the scenarios under which the substance may be harmful. This information is used by technical professionals to establish specialized work site processes for exposure management, preservation, handling, firefighting, spillage cleanup, and so on.

4.7.10 Reactivity and Stability

This part of the SDS explains any situations that may cause the material to become unstable or to react violently, as well as those that ought to be prevented. Unstable compounds can decompose (break down) and create explosions or fires, or they might generate new substances with distinct risks. Factors such as heat, sunshine, and chemical ageing can cause unstable compounds to degrade. When incompatible elements are combined or brought together, they might react violently or explosively. These substances should be maintained separately and not combined unless additional precautions are taken. The details provided in this part must be understood so that the substance may be stored and handled securely and that incompatible materials are not mixed.

4.7.11 Information of Toxicological

This part of the SDS includes toxicity data for the product's constituents or the entire product. This data might be quite complex and difficult to understand. It is used to assist in identifying risks.

4.7.12 Ecological Information

This section offers data that may be used to assess the ecological effect of the item if it is released (for example, danger to fish, plants, birds, and microbes). This material is primarily meant for environmental specialists.

4.7.13 Considerations for Disposal

This area of the SDS is primarily designed for environmental specialists. Generally, general trash disposal information will be supplied. Typically, the SDS frequently does not include the federal, provincial, or municipal requirements that must be obeyed. This information should be obtained from the proper authorities in your region.

4.7.14 Information for Transportation

This portion of the SDS is for individuals who are in charge of transporting the material. If specific measures are required during transportation, then they will be provided. Transportation of Dangerous Goods and product identification number are inscribed in many materials.

4.7.15 Monitoring Information

The part conveys information related to the regulations for workers. Referrals to relevant safety, wellness, and environmental regulations and laws, as well as information on the product's regulatory status, may be included.

4.7.16 Additional Information

Supplemental information that the SDS author deems necessary for the material's safe use (e.g., label wording, hazard classifications) is present in this part. The sources of reference used in the preparation of the SDS are occasionally cited. The date that the SDS was created (or the most recent time it was reviewed or changed) should be specified. When additional details become available, the SDS will be updated.

4.8 When a SDS is Obtained, What Should be Done?

- Make sure there are 16 headers.
- Ensure that it is written in English and that it is clear and simple.
- Ensure that it is up to date and that any alterations and revision information are given.
- Check that the information on the chemical's label matches the information in Section 2 of the SDS.
- If you are dissatisfied with the information given, reach out to the supplier and seek an updated version.
- Use the data to create chemical risk assessments, teach staff about the risks of the chemical, precautionary precautions to take when employing it, and emergency procedures.
- Keep the SDS in a location that is known and readily available to all personnel, either as a paper copy or electronically.
- Maintain a chemical catalogue of all chemicals on the spot, and SDSs are a great tool for keeping track of all compounds in the workplace.
- It is also a good idea to verify the SDSs on a regular basis to ensure that duplicate or outdated documents are not kept.
- If an exposure scenario is appended to the SDS, then make sure the risk control measures related to the substance's usage are in place.

Bibliography

Ronald, J.W., Understanding a safety data sheet (SDS) in regards to process safety. *Procedia Eng.*, 45, 857–867, 2012.

Bernstein, J.A., Material safety data sheets: are they reliable in identifying human hazards? *J. Allergy Clin. Immunol.*, 110, 1, 35–38, 2002.

Nicol, A.M., Hurrell, A.C., Wahyuni, D., McDowall, W., Chu, W., Accuracy, comprehensibility, and use of material safety data sheets: a review. *Am. J. Ind. Med.*, 51, 11, 861–876, 2008.

Kolchinski, A.G., When safety data sheets are a safety hazard. *Org. Process Res. Dev.*, 26, 2, 263–266, 2022.

Eastlake, A., Hodson, L., Geraci, C., Crawford, C., A critical evaluation of material safety data sheets (MSDSs) for engineered nanomaterials. *J. Chem. Health Saf.*, 19, 5, 1–8, 2012.

Greenberg, M.I., Cone, D.C., Roberts, J.R., Material safety data sheet: A useful resource for the emergency physician. *Ann. Emergency Med.*, 27, 3, 347–352, 1996.

5

Safety in Chemical Laboratories in Academic Institutions

Abstract
This chapter discusses the entry-level safety protocols followed in a chemical laboratory at an institution. The RAMP concept was discussed in detail, starting from dealing with a chemical hazard to its prevention. Lastly, incident prevention by following a few safety measures is also discussed in detail.

***Keywords*:** Safety, chemical hazard, PPE, waste disposal

5.1 Background

A chemical laboratory is an essential part of any university. As a result, everybody working in such a lab must have a firm foundation in the fundamental concepts and processes of laboratory safety and enhance their expertise. A person working in the chemical laboratory should be concerned about safety and cultivate a safety ethic in order to spot safety concerns and make decisions that represent their personal safety beliefs and knowledge. As a result, it is critical to follow and improve safety rules in order to create awareness of chemical safety principles and enable employees to use these ideas when working in a laboratory.

5.2 Introduction

As compared to other science laboratories, chemistry laboratories pose higher risks. Surprisingly, the features that humans love in certain compounds are also the ones that make them dangerous. Humans like the fact that certain organic solvents dissolve molecules of organic material extremely well, yet this same property causes them to dry out our skin. We exploit acids and bases' reactivity to produce a chemical shift; however,

Abhishek Barua, Kanchan Kumari, Sumit Saha, Siddharth Jeet and Swastik Pradhan. Institutional and Industrial Safety Engineering Practices, (73–82) © 2025 Scrivener Publishing LLC

their reactivity also makes them dangerous if they come into contact with skin or are consumed. People assume that liquid nitrogen and dry ice are equally incredibly cold because they need them at times, yet these chemicals are unsafe to handle with bare hands because they are so cold. They know that natural gas ignites, but, if it accumulates in a confined space, then flames or sparks might create an explosion. In order to work safely in laboratory, people need to learn lots of rules about chemicals and how to handle them. There are principles of safety behind all of these rules which deal with many situations. The safety knowledge and skills that one learns and follows are greatly influenced by the safety culture of their institution.

5.3 The RAMP Concept

RAMP is a scientific safety concept created by American chemists Robert H. Hill and David C. Finster [1]. The authors provide RAMP as a memo to help practitioners, educators, and students recall four fundamental safety principles and the major learning objectives associated with each principle. The RAMP concept provides a framework for making chemical safety an integral part of every workplace. RAMP is an acronym that stands for: Recognize, Assess, Minimize, and Prepare.

- Recognize the hazards: One can recognize the hazard by first understanding its nature and identifying the most important physical hazards from equipment, conditions, and procedures. When they know this, they can place it in a category such as flammable, corrosive, explosive, oxidizer, toxicant, or irritant. Have familiarity with hazard rating systems and hazard identification (e.g., the Globally Harmonized System). Be able to explain these more detailed learning objectives on how to recognize hazards. Generally, there are four categories of hazards associated with chemical laboratories (Figure 5.1).
- Assess the hazard risk: One can accurately assess the risks of specific hazards by considering how they might be exposed and what the result of the exposure would be or by creating their own experiment. The possibility of being injured or harmed as a result of being exposed to hazardous substances, processes, or equipment is referred to as risk. The capacity to evaluate the corresponding severity of a certain process' dangers and offer an exact likelihood of being subjected to

Physical Hazards	Chemical Hazards	Reaction Hazards	Health Hazards
• Compressed gases and high-pressure systems • Cryogenic hazards • Electrical hazards • Incompatible chemicals for storage or handling • Low-pressure and vacuum systems • Radiation hazards	• Corrosive chemicals • Flammable chemicals	• Catalyst effects on reactions • Peroxides and peroxide-forming chemicals • Reactive and unstable chemicals • Scale-up and potential runaway reactions	• Acute chemical exposures • Chronic chemical exposures • Nanomaterials • Toxic substance classifications

Figure 5.1 Common laboratory hazards.

such hazards within specified conditions is required. They must be capable of deciding whether the likelihood of exposure or the severity of the threat is more essential in a given case.

- Minimize the hazard risk: The ability to recognize the reduction of risk level can be done by applying proper engineering controls, organizational measures (training, techniques, processes, and policies), and personal protective gears is required. Hazardous risks can be reduced by employing appropriate instruments, approaches, and processes to safeguard investigators as well as others throughout laboratory work. Laboratory hoods, restraints, shielding, interlocks, standard operating procedures, other equipment, and suitable PPE should all be used to reduce exposure to dangerous substances. Through personal inquiry or conversation with the research principle investigator, one should be able to clarify the fundamentals of these protective measures as well as the benefits and drawbacks of various approaches. It is important to understand the typical strategies for minimizing danger as well as the constraints of those precautions. It should be understood that there are dangers in any laboratory research and that thorough preparation and planning are essential to reducing them.
- Preparation for emergency: It goes without saying that you should always be prepared. Whether it is an accidental spill or maybe even a fire, you should be aware and knowledgeable about how to contain any accidents. Therefore, have

a plan in place and know how to execute it. They should understand the value of training, know the locations of all emergency equipment, and consider what one would do in the event of an emergency.

5.4 Incident Prevention

The prevention of any mishap is a collaborative duty that requires everybody in the laboratory to work together [2]. Although everyone in the laboratory is accountable for safety, the experimenter has the greatest direct control over occurrences. Inadequate mindset regarding safety, failure to recognize risks or hazardous situations, inability to properly evaluate the risks associated with the job that is being done, failing to be aware of one's surroundings, failing to comply with guidelines or measures to minimize risks, and failure to recognize the limitations of one's knowledge and experience are all common causes of incidents. Anyone can become a victim of their own or someone else's error. If they are doing a laboratory technique poorly and a lab mate flags this to you, then thank them because he or she may have just saved a life. Inform a labmate if you see them making a mistake. Furthermore, risky behaviors should be addressed to superiors so that mistakes are not repeated [3]. Some of the important protective equipment everyone working in the chemical laboratory gears up with is as follows:

5.4.1 Personal Protective Equipment (PPE)

Personal protection equipment represents one of the most important ways to keep someone safe when working in a laboratory. PPE is intended to avoid or reduce exposure to certain risks encountered while working in a chemical laboratory, and it comprises devices meant to protect certain portions of the body, such as the hands and eyes. Eye protection goggles, gloves, aprons, and laboratory coats are frequent [4, 5]. One should not rely exclusively on PPE for shielding themselves because it is frequently the last line of defense among them and the exposures.

5.4.2 Dressing for the Laboratory (Hair and Apparel)

Clothing worn in the lab should provide basic splash and spill protection for the skin. Short skirts, shorts, and shirts that expose the skin to possible spillage needlessly. It is usually wise to limit the quantity of skin that gets exposed to laboratory conditions. Bulky and baggy clothes are not permitted in the laboratory. Loose sleeves can knock over laboratory equipment, drag through chemical spills, and pose a fire danger with open flames. Natural fibers, such as cotton, should be used to make clothing. Nonflammable, nonporous laboratory jackets or aprons provide the most effective and cost-effective protection.

Wear shoes with leather or a polymer leather substitute upper that entirely protect the feet and toes in the laboratory to provide maximum protection against spills and fallen goods. When selecting laboratory footwear, keep in mind that the shoes should not show the tops of the feet and should provide balance for walking and standing.

Restriction of long hair and loose attire. Long hair is readily entangled in equipment, can be contaminated by chemicals, and can catch fire when directly exposed to lighted Bunsen burners. Jewelry, such as bracelets, rings, necklaces, and wrist watches, should not be worn in the laboratory. Chemical fumes and vapors, as well as spills, can cause harm to jewelry. Chemical seepage within jewelry and the skin can bring corrosives into close contact with the skin, trapping the chemicals. Jewelry can also get caught on machines and cause harm.

5.4.3 Eye Protection

While not carrying out a chemical operation, everybody in the lab, including visitors, needs to wear eye protection at all times. Some investigations involve splash dangers, necessitating the use of indirectly vented goggles; for others, safety glasses may suffice. Normal prescription eyeglasses lack enough laboratory eye protection against fragmentation from an explosion or chemical spills. Wearing standard prescription eyeglasses without chemical splash goggles or safety glasses has resulted in serious injuries. Eye protection is best provided by safety goggles approved for chemical splash protection.

5.4.4 Gloves

Gloves are an essential component of personal protection. The material of the gloves must be chosen depending on the chemical substances being used. Gloves should be inspected before each use to verify that there are no cracks or tiny holes. Gloves must be taken off before leaving the work area as well as before handling certain items such as mobile phones, computers, writing tools, textbooks, laboratory notebooks, calculators, and doorknobs to avoid accidentally spreading chemicals. Even if wearing gloves, one must wash their hands upon exiting the laboratory. There are many different types of gloves and materials available, including butyl rubber, neoprene, and a few more. Varieties of gloves have varied gauntlet lengths; some simply cover the forearm, others the full arm, and yet others only cover the wrist. Latex-sensitive individuals should avoid wearing latex gloves. Although fabric or leather gloves can guard against both cold and hot things, they cannot protect against harmful substances. Disposable gloves and gloves that have been chemically penetrated should not be reused. Because the chemical cannot be completely removed, the gloves cannot be safely reused. Contaminated gloves are often deemed hazardous trash; however, this is occasionally the case. Used gloves should always be disposed of in an authorized hazardous waste receptacle or as ordered by superiors.

5.5 Protocols for Laboratory

5.5.1 Environment of Laboratory

The chemistry laboratory may offer an endless number of learning possibilities; nevertheless, when working in the lab, everyone should be aware of their own activities as well as the actions of others around them. Procedure modifications, such as alterations to the chemicals to be utilized or the amounts to be utilized, might be hazardous. Only with the knowledge and agreement of the lab supervisor may changes be made. Before beginning work in the laboratory, make a mental note of your surroundings. Exit locations, eyewash fountains, fire alarm pull stations, safety showers, first-aid kits, fire blankets, and fire extinguishers should be known to everyone. Eating or drinking in the laboratory should be avoided to eliminate the possibility of contamination leading to the intake of a laboratory chemical. There should be no beverages or food brought to or kept within the lab space.

5.5.2 Visitors in the Laboratory

All laboratory visitors, regardless of the duration of their stay, should wear appropriate protective equipment because they may be unaware of the dangers and may unwittingly engage in harmful activities. To avoid catastrophes, both visitors and lab personnel should follow proper safety standards.

5.5.3 Housekeeping

Keeping things clean and neat in the laboratory and everywhere else often leads to a safer atmosphere. Keep boxes, chairs, open drawers, garbage receptacles, and backpacks out of aisles and out of the way of safety equipment. To reduce slipping dangers, keep the floor clean of spilt beverages, stoppers, ice, rods or glass beads, and other tiny things. Keep leftover chemicals and broken or soiled glassware out of workplaces and storage rooms. Broken glassware should be discarded exclusively in a broken glass disposal container. After laboratory experiments, the working space should be washed out so that others do not come into contact with chemical residue. Chemicals should not be left on scales because doing so could expose another user to that chemical needlessly. Furthermore, electronic balances are costly and readily destroyed by corrosive substances.

5.5.4 Labeling Chemicals

Numerous mishaps have occurred as a result of improper or inadequate labeling of chemical containers. A manufacturer's label should never be modified, covered, or altered in any way until the container has been proven to be empty. Empty containers are frequently used again, for example, for student-prepared solutions. Before reusing a container for another solution, the outmoded label should be fully removed, and the container must be well cleaned and allowed to air dry. It is not allowed to write over a manufacturer's label using a marker. A container shouldn't ever have two labels, preferably on both sides of the bottle. A secondary label for short-term usage should include the chemical's name, the name of the person who filled the container, when it was filled, and the dangers. Containers designed for long-term storage should be labeled in accordance with the Globally Harmonized System of Classification and Labeling of Chemicals.

5.5.5 Cleaning Glassware

Hot water, ecologically acceptable cleaning chemicals, and brushes of appropriate stiffness and size should be used to clean dirty glassware at the laboratory sink. It is not advisable to press a brush into glassware. When washing dishes, chemical splash goggles and gloves should be used, as directed. Many laboratory faucets feature serrated nozzles capable of producing high-velocity water streams. Always regulate the stream carefully while using these sorts of faucets without holding glasses beneath. Water might splatter back toward your face or knock the glasses out of your hands. It is best to avoid accumulating too many stuff in the cleaning area. The workspace surrounding a washbasin is typically restricted, and stacking dirty or cleaned glassware might result in damage. The sharp, serrated edge of a piece of shattered glassware that was unbroken when it was placed in the water may be hidden by turbid water in a washbasin. Water that is stagnant should be emptied if the glassware in the washbasin is shattered. To remove the shattered glass, use a pair of cut-resistant gloves, tweezers, or tongs. When clearing the drain area, be cautious because glass fragments can become lodged in the perforations and become difficult to detect. Sink bottoms with non-clogging rubber or plastic mats may be used to reduce glassware breakage.

5.5.6 Inhaling Harmful Chemicals

When asked to smell anything in the laboratory, one should use their hand to gently waft vapors toward their face and nose. Never smell a chemical by pressing one's nose against a chemical container. The existence of an odor does not reliably indicate the presence of potential harm, and the absence of an odor does not reliably indicate the absence of danger. Some individuals believe that if they detect a chemical, it is harmful to them. This is not always the case. When people smell a chemical, they are almost certainly breathing it. Some dangerous substances have no aroma, while others have the ability to paralyze one's sense of smell. Some compounds cannot be recognized by the human nose at lethal doses, whereas others, despite having a definitely foul fragrance, are not harmful if breathed in. Many compounds, whether they have an odor or not, are hazardous if their gases, particles, or mists are breathed in. The container label and the safety data sheet for the chemical may both contain a warning regarding inhalation dangers.

5.5.7 Disposal of Chemicals

An effective control of reaction by-products, excess and waste chemicals, and contaminated materials is a critical component of incident prevention, and there are very severe requirements for disposing of chemicals. Improper disposal can cause major environmental damage as well as legal problems for any organization. Every lab worker has responsibility for making sure these wastes are dealt with in a way that minimizes personal risk while also considering the possibility of environmental contamination. Typically, reaction by-products and excess chemicals are dumped into adequately labeled garbage or hazardous waste bins for safe disposal.

Various containers will most likely be used for various groups of chemicals. Waste items should be disposed of in the manner specified by the lab instructor. Pouring garbage into an improper container may cause unanticipated negative responses, such as fires or explosions. Sometimes, reaction by-products can be neutralized or deactivated as a component of the method, which can assist in decreasing waste handling and hence cutting disposal costs.

The following are some standard disposal instructions:

- While disposing of chemicals, every kind of discarded chemical should be placed in its own clearly designated disposal container. The information on the label should be thoroughly read, and the cap that seals it should be replaced after use.
- Chemicals should not be put into a sink or down the drain, which can contaminate water bodies indirectly.
- Separate ordinary wastepaper from chemical waste in a wastepaper basket. Chemically contaminated materials, such as tissue paper used to wipe up a spill, might require storage in a specific container labeled for this purpose.
- Broken glass should be placed in its own clearly labeled trash container. If the shattered glass contains chemicals, then it must be disposed of properly in an appropriate disposal bin.

References

1. Finster, D.C., RAMP: A safety tool for chemists and chemistry students. *J. Chem. Educ.*, 98, 1, 19–24, 2020.
2. Walters, A.U., Lawrence, W., Jalsa, N.K., Chemical laboratory safety awareness, attitudes and practices of tertiary students. *Saf. Sci.*, 96, 161–171, 2017.
3. Huston, E.M., Milligan, J.A., Powell, J.R., Smith, A.M., Neal, D., Duval, K.M., DiNardo, M.A., Stoddard, C., Bell, P.A., Berning, A.W., Wipf, P., Development of an undergraduate course in chemical laboratory safety through an academic/industrial collaboration. *J. Chem. Educ.*, 95, 4, 577–583, 2014.
4. Reniers, G.L., Ponnet, K., Kempeneers, A., Higher education chemical lab safety interventions: Efficacious or ineffective? *J. Chem. Health Saf.*, 21, 1, 4–8, 2014.
5. Pesta, S. and Kaufman, J.A., Laboratory safety in academic institutions. *J. Chem. Educ.*, 63, 10, A242, 1986.

6

First Aid and Compressed Gas Safety in Academic Institution Laboratories

Abstract

This chapter discusses the importance of lab safety and the knowledge of first aid necessary in a chemical laboratory at an institution. Prevention of lab accidents, preparation, and providing first aid during any accident have been discussed. The latter section discusses compressed gas cylinder handling and related safety. Different categories of compressed gases used in institutions, along with the Standard operating procedures for handling them, have also been stated.

Keywords: First aid, chemical hazard, compressed gas safety

6.1 Background

First aid is critical in labs for protecting the safety and well-being of those who work with potentially dangerous products and equipment. Safety in the laboratory is a shared responsibility. Regular safety training, attention to protocols, and knowledge of possible risks are critical to preventing accidents and efficiently responding in the event of an emergency. Compressed gas cylinders are widely used in laboratories for a variety of functions, including delivering gases for experiments and running equipment. To avoid accidents and possible risks, compressed gases must be handled and stored carefully. When dealing with compressed gases, it is critical to study appropriate safety recommendations, legislation, and your organization's unique policies. Each type of gas may have unique safety implications; therefore, it is critical to become acquainted with the qualities and risks of the gases with which one is dealing [1].

Abhishek Barua, Kanchan Kumari, Sumit Saha, Siddharth Jeet and Swastik Pradhan. Institutional and Industrial Safety Engineering Practices, (83–100) © 2025 Scrivener Publishing LLC

6.2 Introduction to First Aid at Laboratories

A laboratory can be dangerous owing to the existence of caustic liquids, hazardous chemicals, toxic fumes, and poisonous compounds. However, no one should be alarmed because the risks can be minimized by using laboratory safety precautions. First aid is a set of principles that can lower the damage caused by injury or chemical exposure before expert medical care can be delivered. Laboratory first aid is the urgent medical help given to those who are wounded or have health-related occurrences in a laboratory environment. Because laboratory work involves chemicals, equipment, and potentially dangerous substances, it is critical to be familiar with laboratory-specific first aid methods. First-aid protocols should be familiar to everyone in the laboratory because they save time and reduce the extent of injury in the case of a laboratory tragedy [2].

6.2.1 Why is Lab Safety and Knowing First Aid Necessary?

One may not realize it, but a scientific lab, whether in a classroom or professional environment, is quite dangerous if basic safety precautions are not taken. Working with glass equipment, live animals, chemicals, and even plain water can result in catastrophes such as fires, chemical burns, slips and falls, animal bites, cuts from shattered glass, and toxic fume inhalation. All of this is conceivable in settings where safety rules are not properly enforced. Accidents may and do occur even in the safest laboratories. That is why, in addition to lab safety requirements, one should know first aid to address these injuries and limit their consequences until first responders arrive or expert medical assistance is provided [3].

6.2.2 Preventing Lab Accidents at Institution

Dealing in a lab requires being completely aware of what one is dealing with and the nuances of the task at hand. Mostly pupils may be unfamiliar with dealing with chemicals or in a lab along with the other persons who should be aware of the risks or hazards associated to the work they are doing. When doing a new activity or operation, everyone should familiarize themselves with the directions and specifications beforehand [4].

6.2.2.1 Preparation

Before someone begins working, they should ensure that they are fully prepared. This includes the following:

- Examining equipment and containers for flaws such as cracks or chipping.
- Examining electrical equipment for flaws like frayed cables or exposed wires.
- Locate the emergency eyewash stations.
- A first-aid kit should be appropriately supplied.
- Removal of any debris and looking for any tripping risks.
- Never work in a laboratory by yourself.

6.2.2.2 Clothing

Dress for safety success by adhering to the following clothing and apparel guidelines:

- Protect the arms and legs by wearing correctly fitting long sleeves and long trousers or skirts.
- Put on sturdy closed-toed shoes with flat heels. High heels are a tripping danger, whereas woven fabrics and sandals expose your flesh.
- Tying of long hair at the back.
- Taking off any jewelry.
- Wearing Proper protective gears while using any chemicals.

6.2.2.3 Prevention of Contamination

Working in a lab requires one to avoid cross-contamination. Some of the important points are as follows:

- Never bring food or beverages inside the lab. To ensure safety, keep lip balm, lotions, and other cosmetics away from the lab.
- Remove gloves and properly dispose of them in a designated location followed by washing of hands with soap and water.
- Handling sharps with care (syringes and needles), and dispose of them in an approved container.
- Keep items well contained, especially when taking them out of the lab.
- Use of clean goggles, lab coat, gloves, shoes, and mask.

6.2.2.4 *Care with Chemicals*

- Avoid fumes by working in a good ventilated space and under hood for chemical fume.
- Avoid smelling, inhaling, or tasting chemicals.
- Never flush chemicals down the sewage; it should be placed in designated area.
- When not in use, make sure that lids are tight and containers are secure.

6.2.3 Preparation of First Aid for Lab Accidents

Laboratory surroundings may be hazardous, and it is critical to be prepared for such mishaps or injuries. The following are some laboratory first aid essentials to remember:

- Assess the situation: Put safety first by assessing the scenario before approaching the wounded individual. Identify any continuing dangers or risks and adopt the necessary mitigation steps. Evacuate the area if required to protect yourself and others.
- Basic first aid instruction: Learn basic first aid, including Automated External Defibrillator (AED) and Cardiopulmonary Resuscitation (CPR) use. This understanding will enable you to give quick aid in life-threatening circumstances.
- Knowledge of emergency procedures: Become acquainted with the laboratory's emergency protocols and locate emergency exits, eyewash stations, safety showers, fire extinguishers, and other safety equipment. In the event of an emergency, follow these steps.
- First-aid kit: Every laboratory should keep a well-stocked first-aid kit on hand. Items such as sticky bandages, sterile gauze pads, antiseptic solutions, adhesive tape, gloves, scissors, and tweezers should be included. Check the kit on a regular basis to verify that all supplies are up to date and well packed. Ensure that emergency equipment such as eyewash stations, safety showers, and fire extinguishers is in good working order and that their locations are properly indicated.

- Personal protective equipment (PPE): When working in the laboratory, always use proper PPE such as gloves, safety goggles, lab coats, and closed-toe shoes. This safeguards you from potential injuries and chemical exposures.
- Communication: Establish clear communication channels inside the laboratory. Ascertain that all lab members are aware of the location of emergency contact numbers, first-aid kits, and first aid professionals.
- Documentation and reporting: Keep detailed records of any events or injuries that occur in the lab. Report the occurrence to the appropriate people or departments within your organization so that suitable follow-up steps may be performed to prevent similar events in the future.
- Risk assessment: Before beginning any laboratory work, conduct a risk assessment. Determine possible hazards and take the required actions to reduce the risks. This may involve the use of fume hoods, careful chemical handling, and adequate ventilation.

6.2.4 First Aid for Lab Accidents

Accidents can happen even when all safety precautions are in place, which is why it is critical to have a well-stocked first aid bag and understand first aid in lab settings. Here are some important considerations for laboratory first aid:

- Seek assistance: If the injury is serious or life-threatening, then contact emergency medical services or the official emergency number in your region right away. Give detailed details regarding the circumstances, location, and kind of injury.
- Take precautions: Before assisting the wounded individual, put on personal protective equipment (PPE) such as gloves and safety goggles. This guarantees that you are protected from any potential risks or pollution.
- Injuries and wounds: Labs rely on glass pipettes, beakers, and other containers, which can break and cause injuries. If you are with someone who has a minor cut, then apply sterile gauze pressure to the wound. Apply a sterile bandage after washing with soap and water. For more severe cut, have the person sit down and lift the bleeding section while

applying pressure. Continue to apply pressure while driving them to an urgent care or emergency facility. If the bleeding is significant or the person feels weak or lightheaded, then contact the nearest hospital's medical team or ambulance.

- Stop bleeding: If bleeding occurs, then direct pressure should be applied to the area with a clean towel or sterile bandage. If possible, elevate the damaged region to help limit blood flow. Seek expert medical care if the bleeding is serious and uncontrollable.
- Substance exposures: If someone comes into contact with a dangerous substance, then remove them from the exposure source as soon as possible. Rinse the affected area thoroughly with water for the time required by safety standards or chemical safety data sheets (SDS). If required, encourage the person to flush their eyes. Seek medical assistance as soon as possible and show the SDS or any information regarding the chemical to the healthcare professional.
- Burns: Different forms of burns, such as thermal, chemical, or electrical burns, can occur in a laboratory. Cool the injured region for at least 10–20 min with cold water if there is a thermal burn. Take off any clothes or jewelry that is not stuck to the burn. Apply no ointments or lotions to the burn. Apply a sterile, nonstick dressing to the burn. If you have chemical burns, then rinse the afflicted area with water for the required amount of time and seek medical assistance. Before approaching an injured individual with electrical burns, make sure the power source is turned off. Cover the burn with a sterile dressing and seek medical attention right away.
- Eye injuries: Do not touch or massage someone's eye if they have an eye injury. Encourage the person to keep their eyes open for at least 15 min and to gently rinse them with clean water or a sterile eyewash solution. If an item becomes lodged in the eye, then do not try to remove it. Allow water to stream across both eyes while the subject rotates their head from side to side. If you get dust or a solid object in your eye, then cover or shut it and seek medical attention.
- Cardiopulmonary resuscitation (CPR): Ensure that you and your staff are fully educated in first aid in case an accident occurs in your lab! Quick thought and quick action may be the difference between a good and unsuccessful conclusion,

and, when your whole team of students or employees knows how to react, the lab becomes a safer place to learn or work. If someone is unconscious and not breathing in the laboratory, then start CPR if you are trained and certified to do so. Begin with chest compressions and continue the procedures until expert medical assistance arrives.

Maintaining a safe and productive working environment requires laboratory first aid. Regular training, an understanding of laboratory-specific dangers, and a fast reaction to crises can all help reduce the effects of accidents and protect laboratory personnel's safety.

6.3 Introduction to Compressed Gas Cylinder Handling

Throughout the institution, gas cylinders are employed to distribute certain gases. They come in a variety of shapes and types, ranging from basement gas to industrial and laboratory gases (see Figure 6.1). The volume, size, and form of gas cylinders vary greatly. They are all made of highly strong material, generally steel, and are designed to confine gas at high pressures. The full pressure varies based on the type of gas and the duty of the cylinder, and pressures can reach extremely high levels. Cylinders are classified into two types: liquid gases and compressed gases [5].

The cylinders, the gases stored in the cylinders, the piping and connections, and the usage of the gases all pose risks. There are inspection and maintenance requirements for gas supply systems, and information is

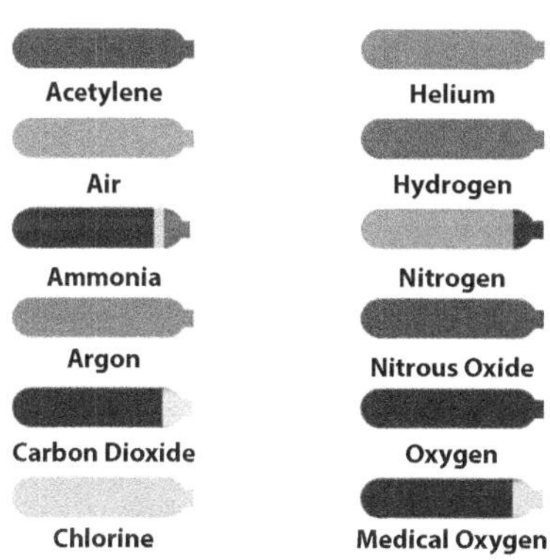

Figure 6.1 Different types of compressed gas cylinder.

provided to help decide what is necessary. Gas cylinders are dangerous in the case of a fire. When deciding how to combat a fire, the Fire Service may conclude that gas cylinders constitute a risk to their personnel [6].

Gas usage must be justified and supported by an appropriate risk assessment. To assist in deciding on additional controls and regulations, a general risk assessment for the introduction of flammable gases into a work environment is offered. Departments that require or use gas cylinders must register with the health and safety department and have plans for procuring, storing, utilizing, and disposing of them.

6.3.1 Different Categories of Compressed Gases Used in Institutions

a) Asphyxiant Gas: Asphyxiant gases are a kind of gas that can displace or lower the concentration of oxygen in the air, resulting in oxygen deprivation and the possibility of suffocating or asphyxiation. Because these gases are colorless, odorless, and tasteless, they are especially harmful because their existence is not always obvious.

b) Compressed Air: While not technically a gas, compressed air is frequently employed in a variety of applications. It is primarily produced by compressing ambient air and can be used for cleaning, pneumatic equipment, and power generation. However, compressed air may be dangerous if it is mishandled or contaminated.

c) Compressed medical gases: These gases comprise gaseous and liquid (cryogenic) forms that are held in high-pressure cylinders and supplied as a gas.

d) Corrosive gases: These are extremely reactive and can cause harm to biological tissues, equipment, and buildings. On contact, they may corrode metals, damage materials, and pose serious health concerns. Hydrogen chloride, sulfur dioxide, and chlorine are a few examples. When working with corrosive gases, certain handling and storage measures, including the use of suitable personal protection equipment, are required.

e) Cryogenic Gas: Any gas having a boiling point less than −150°C is considered a cryogenic gas. Oxygen and nitrogen are the most prevalent cryogenic gas.

f) Flammable Gases: In the presence of an ignition source, these gases have the ability to ignite and burn. Hydrogen, methane, propane, butane, and acetylene are a few examples. To avoid fire and explosion threats, flammable gases must be stored, handled, and transported with extreme care.
g) Inert Gases: Because inert gases do not readily react with other chemicals, they are valuable in applications requiring an unreactive environment. Nitrogen, argon, helium, and carbon dioxide are a few examples. In welding and other industrial operations, inert gases are frequently used for purging, blanketing, and shielding.
h) Oxidizing Gases: Oxidizing gases aid combustion and can start or exacerbate fires. They emit oxygen or other oxidizing chemicals that can violently react with combustible things. Oxygen, nitrous oxide, and chlorine are a few examples. To avoid the inadvertent ignition of combustible materials in the presence of oxidizing gases, proper handling and storage practices are essential.
i) Toxic gases are detrimental to human health when breathed or applied to the skin. They can cause everything from discomfort and respiratory difficulties to serious poisoning and death. Ammonia, chlorine, hydrogen sulfide, and carbon monoxide are a few examples. When working with dangerous gases, strict controls and sufficient ventilation are required [7].

6.3.2 Standard Operating Procedures for Handling Compressed Gases

6.3.2.1 General Requirements

- Check the label for the correct contents of cylinder before use. Do not use color coding as a sure method of identification. No labels, tags, or stencil marks used by the seller to identify the contents should be defaced or removed.
- When cylinders are not in use, protective caps must be retained on them. The cap on the cylinder neck must be manually tightened all the way down. The word "in use" refers to being connected to or actively using a piece of equipment.

- To avoid falling, cylinders must be adequately fastened at all times.
- Any evident damage that might compromise the cylinder's integrity, such as rust, should be brought to the vendor's attention while returning the cylinder.
- The contents of one compressed gas cylinder should never be transferred to another cylinder.
- Evacuate the room and inform the Safety Department if any cylinder is suspected of leaking. Notify the supplier of the damaged cylinder as soon as possible.
- If a cylinder holding dangerous gas leaks, then vacate the room immediately, lock the door(s), activate the fire alarm, escape the zone, and telephone the Safety Department from a safe position to report the situation. Meet with the emergency personnel to offer any further information that may be relevant.
- Repeat the preceding procedures for a cylinder holding flammable or oxidizing gas. Before leaving the location, turn off any ignition sources if they are easily accessible and can be done safely.

6.3.2.2 *Transportation of Cylinders*

- When transporting cylinders, protective valve caps must be fastened. Never move or raise the cylinder by its cap.
- Move gas cylinders with a hand truck or equivalent gear (see Figure 6.2). Never pull or roll a cylinder, and avoid collision of a cylinder with other surface or cylinder.
- When moving or putting any cylinder, always wear gloves and eye protection.
- Never suspend cylinders using ropes or slings unless the vendor provides connection points on the cylinder.

6.3.2.3 *Storage of Cylinders*

- Storage spaces for cylinder must be clearly labeled with the danger class or the gas name. "Flammable Gas Storage Area" and "No Smoking" are some examples.
- Anchor upright gas cylinders with valve end up, to a wall.

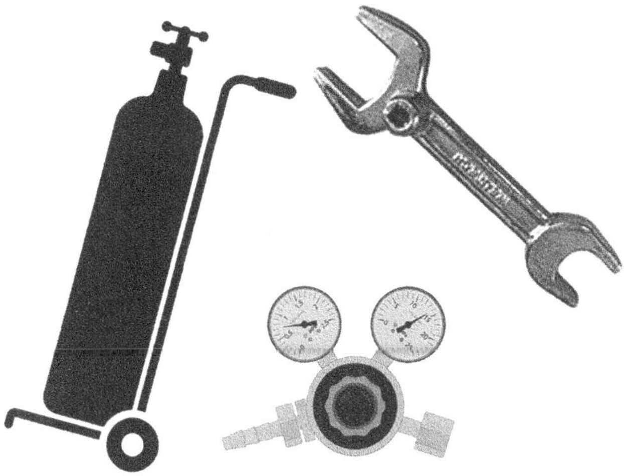

Figure 6.2 Compressed gas cylinder trolley, pressure regulator, and wrench.

- Never keep cylinders in stairwells, lift lobbies, hallways, or any other position that might hinder the building's residents' safe departure pathway.
- If multiple types of gases are kept in the same area, then "full" cylinders must be separated from "empties." Other types of segregation include grouping cylinders by gas type, such as oxidizing, corrosive, or flammable.
- Store cylinders in a well-ventilated place away from any source of ignition or heat.
- Cylinders should kept away from any salt or other corrosive substances or exposure to persistent wetness. Cylinder valves can be damaged by corrosion. Corrosive chemical cylinders must be examined on a regular basis to verify that the valve has not rusted.

6.3.2.4 *Use of Compressed Gas Cylinders*

Refer to all the label information and SDS linked to the gas before using cylinders. Follow these instructions to set up and use the cylinder:

- Connect the cylinder to the closed regulator. If the regulator is not entirely closed, then never open the cylinder valve. A regulator should be attached to a cylinder without putting any strain on the threads. It should not be loose either.

- Counter-clockwise spin the screw for delivery pressure adjustment until it readily rotates.
- Slowly open the cylinder until the cylinder pressure is detected by the regulator's intake gauge.
- Turn the screw clockwise for delivery pressure adjustment until the appropriate pressure is achieved.
- Use soap solution or a Snoop to check for leaks. To identify leaks, never use an open flame.
- After using the gas, cylinder valve should be closed and regulator pressure should be released followed by reinstalling the gas cap.

6.3.2.5 *Piping and Equipment Assembly*

- Never push threads that do not fit perfectly.
- For assembling, use thread lubricant or Teflon tape.
- Avoid making abrupt curves in copper tubing it make it harden or break.
- Inspect tubing on a regular basis and replace it as needed.
- Most pressure work does not require Tygon or plastic tubing. These materials can fail when subjected to pressure or temperature stress.
- Copper pipe should not be used for acetylene.
- For chlorine, do not use cast-iron plumbing.

6.3.2.6 *Leakage in Cylinders*

The majority of leaks happen near the top of the cylinder at the valve, which may include the valve threads, valve stem, valve outlet, or pressure relief mechanisms. It is not safe for lab employees to attempt to fix leaky cylinders. Some of the important instruction followed at that time:

- Transfer the cylinder to a well-ventilated, isolated place (combustible items should not be near it in case of an oxidizing or flammable gas).
- When a significant or unmanageable leak develops, leave the space and alert safety department immediately.

6.3.2.7 *Handling of Empty/Used Cylinders*

- Take off the regulator followed by covering the top with the cylinder cap.
- Labeling the cylinder "empty" or "MT." Keep it in a dedicated spot until it can be returned to the supplier.
- Empty and full cylinders should not be kept together.
- Empty and full cylinders should not be connected to the same manifold because reverse flow can occur toward the empty one.
- Never refill empty cylinders. Gases should only be refilled by the cylinder provider.

6.3.2.8 *Precautions for General Use*

- Never use a wrench on a cylinder valve with a hand wheel. A malfunctioning valve should be marked and brought to the attention of the supplier.
- Do not tamper with, repair, or modify valves, cylinders, or other safety relief systems. For any repairs, notify the vendor and return the cylinders.
- Never use a lever in the cap ports to remove a jammed cap. When the cap is turned, the lever may mistakenly open the valve.
- Do not install cylinders in areas where they might become part of an electric circuit or in touch with an electrically energized system.
- Check the gas transmission system for leaks using "Snoops," soapy water, or leak-detecting equipment. To test oxygen connections, "Snoops" should not be used.
- Make use of pressure regulators having integrated pressure release systems.
- Never let grease, oil, or other flammable substance fall into contact with cylinders and valves, particularly oxygen cylinders.
- Labeling of all gas lines going from a gas cylinder must be properly for indicating the conveyed gas.

6.3.2.9 *Compressed Gas Utilization*

- All label information, along with the SDS, should be followed before using any gas.
- Check that the tank is securely fastened to the proper structure or stand.
- Before connecting a regulator to a cylinder, always inspect it. Always use the right regulator for the gas in the cylinder.
- Ensure that the threads on the connection and the cylinder match before connecting them, and are of the gas-service type.
- The threads and mating surfaces should be cleaned before attaching the regulator and hose, followed by wiping the outlet with a lint-free, dry, clean cloth.
- Fasten the regulator securely with the auxiliary valve closed and the regulator flow set to off (in the anticlockwise direction) before opening the cylinder valve.
- Always use a cylinder wrench for tightening the nut of regulator and tube connectors (see Figure 6.2).
- Teflon tape should not be used on cylinder or connections of tube-fitting.
- Cylinder valves should be opened slowly. Never use a wrench on a cylinder valve with a hand wheel. Return the cylinder to the vendor in case of frozen valve and inoperable by hand.
- Cylinder valve should be closed after used and all the pressure should be removed before removing a regulator from any cylinder.
- Never totally empty a gas cylinder. When the cylinder has at least 30 psi remaining, stop using it.

6.3.2.10 *Precautions for Using Any Flammable Gas*

While handling flammable gases, the following precautions must be followed:

- Flammable gas cylinders (full or empty) should be isolated from oxidizing gas cylinders.
- Keep oxidizing or flammable gases away from solvents or objects that are combustible, gas fires, unsecured electrical connections, or any other ignition source.

- To minimize accumulation of static electricity, the systems dealing with flammable gas should be bonded and grounded.
- While working with or on any flammable gas cylinders/systems, non-sparking equipment must be utilized.
- The amount of combustible gases in a given region should be maintained to a minimum. Other safety equipment, such as a ventilated gas cabinet, may be necessary.
- Never run air tools with compressed oxygen. Such gases must only be utilized for their authorized uses.
- The cylinder valve for acetylene should not be opened for more than one full revolution. Place the valve key or wrench over the valve when it is open so as to close it swiftly in the event of a fire or disaster. Never use acetylene at pressures above 15 psi. The gas release pressure must not exceed the manufacturer's specifications.

6.3.2.11 *Precautions for Using Poison Gases*

When handling dangerous gases, the following precautionary measures should be followed:

- Acutely poisonous gases should be kept in a vented area.
- In a laboratory, the amount of toxic gas used must be maintained to a minimum level.
- Mark and identify places correctly where hazardous or toxic gases are used.

6.3.2.12 *Precautions for Oxygen and Oxidizing Gases*

While working with oxygen or any oxidizing gases, the following precautions must be taken:

- Avoid allowing grease or oil to come into contact with valves or cylinders. Oil and other reducing substances must be removed from regulators and tubes used with oxidizing gases.
- Using "Snoops" to verify oxygen connections is not advised.
- Always slowly open the oxygen tank.
- Keep oxidizing gases away from solvents or objects which are combustible, gas fires, unsecured electrical connections, or any other ignition source.

6.3.2.13 Instructions for Medical Professionals (for Medical Institutions)

- Keep cylinders in designated spaces with clearly labeled signage. Cylinders should not be stored in public locations for example hallways or stairs.
- Secure cylinders correctly on stands or racks.
- The cylinders should be secured and correctly fastened when "in use."
- Oxygen cylinders should not by place on stretchers or patients.
- No excessive medicinal gases are permitted in operating rooms.
- Separate the cylinders into empty and filled ones and name them accordingly. Separate cylinders according to their contents.
- Maintain cylinders in an upright and secure posture at all times.
- Only specially qualified personnel should use anesthetic gas. The employee must have recorded training and be certified by their manager.

Toxic, flammable, oxidizing, corrosive, inert, or a mix of these risks can exist in compressed gases. Furthermore, to the chemical risks, compressed gases can be very pressurized. Because of the quantity of energy in a compressed gas cylinder, it has the potential to be a rocket. It is critical to use caution when handling and storing compressed gas cylinders [8].

References

1. Eisenburger, P. and Safar, P., Life supporting first aid training of the public—review and recommendations. *Resuscitation*, 41, 1, 3–18, 1999.
2. Das, M. and Elzubeir, M., First aid and basic life support skills training early in the medical curriculum: curriculum issues, outcomes, and confidence of students. *Teach. Learn. Med.*, 13, 4, 240–246, 2001.
3. Leonard, L.G., Scheulen, J.J., Munster, A.M., Chemical burns: effect of prompt first aid. *J. Trauma*, 22, 5, 420–423, 1982.
4. Li, F., Jiang, F., Jin, X., Qiu, Y., Shen, X., Pediatric first aid knowledge and attitudes among staff in the preschools of Shanghai, China. *BMC Pediatr.*, 12, 1, 1–7, 2012.

5. Ngai, E. and Ngai, C., Compressed gas safety at the university. *J. Chem. Educ.*, 98, 1, 57–67, 2020.
6. Cadwallader, L.C., Compressed gas safety for experimental fusion facilities. *Fusion Sci. Technol.*, 47, 4, 989–994, 2005.
7. Reinhard, H.F. and Fetherston, F.R., Safe Practices for Handling Compressed Gases. *Ind. Eng. Chem.*, 49, 10, 1751–1754, 1957.
8. Feeley, T.W., Bancroft, M.L., Brooks, R.A., Whyte, J.H., Potential hazards of compressed gas cylinders: A review. *Anesthesiology*, 48, 1, 72–74, 1978.

7

Sharps Safety at Academic Institution Laboratories

Abstract

Ensuring sharps safety at academic institution laboratories is crucial to protecting the health and well-being of researchers, students, and laboratory staff. This chapter contains basic ideas about sharps and hazards related to them. How sharps accidents occur, how to handle them, and the disposal of sharps have been discussed here. In the latter part, the first aid for any sharps accident has been stated.

***Keywords*:** Sharps, disposal, handling, first aid

7.1 Background

Lab apparatus advancement has reduced the use of sharps such as needles, scalpel blades, and glass pipettes over time, but, occasionally, it is unavoidable. One of the most effective ways of transferring infection is through skin penetration with a biologically infected sharp object. All individuals who undertake jobs involving sharps must be careful to minimize the risk of work-related injuries caused by sharp devices. When employing such things, extreme caution must always be exercised. Proper recommendations can help safeguard anyone from suffering sharps-related harm.

7.2 Introduction to Sharps

Needles, scalpels, lancets, metal wire, razor blades, scissors, pins, retractors, staples, clamps, cutters, and glass objects are all examples of "sharps" that may cut the skin. Secondary infections are a danger with all sharps injuries [1–3]. Furthermore, users are exposed to other dangerous or infectious agents through aerosols and sprays while using syringes or needles, mostly when using non-needle-locking syringes. Individuals who deal

Abhishek Barua, Kanchan Kumari, Sumit Saha, Siddharth Jeet and Swastik Pradhan. Institutional and Industrial Safety Engineering Practices, (101–110) © 2025 Scrivener Publishing LLC

with sharps should exercise caution to avoid injury and exposure to biological, chemical, and other potentially hazardous substances.

To minimize autoinoculation and the formation of aerosols while using and disposal, syringes and needles should be handled with caution. Following usage, syringes should be quickly disposed of in a sharps container; exposures from needle sticks have been documented when spent syringes are collected on a work surface for disposal following animal inoculations. Needles should not be sheared, bent, replaced in the needle sheath or guard, or detached from the syringe after use. Sharps containers must be discarded as hazardous waste and appropriately handled before final disposal.

7.3 Hazards Associated with Sharps

Sharps may be dangerous, especially if they are not handled or disposed of appropriately. The following are some of the risks linked with sharps:

- Injuries due to puncture: The greatest concern with sharps is the risk of puncture injuries. Needles, shattered glass, and other sharp items can pierce the skin and underlying tissues, causing sores and infection.
- Transmission of blood borne pathogen: Sharps that have come into contact with blood or body fluids can transfer blood-borne diseases such as the hepatitis B, hepatitis C, and human immunodeficiency virus. If a contaminated needle pierces the skin, then these microorganisms can enter the body and cause severe diseases.
- Allergic reactions: Some people are sensitive to the components included in sharps, such as latex or metals, which can trigger allergic reactions when they come into contact with their skin.
- Risk of infection: Even if a sharp is not contaminated, a puncture wound might nonetheless result in infection. Bacteria from the wound or the surrounding environment can enter and cause localized or systemic illnesses.
- Exposure to chemical: Some sharps used in medical operations, such as shattered glass or needles, may be tainted with toxic chemicals. These compounds can be hazardous if exposed through a puncture wound.

- Entrenched objects: If a sharp object becomes entrenched in the body, it can be difficult to remove and may necessitate specialized medical treatment.
- Secondary injuries: A person may inadvertently harm himself with a sharp while attempting to dispose of it inappropriately or without necessary care in some situations.
- Impact on environmental: Improper sharps disposal can endanger the environment and waste management staff. Sharps that wind up in common trash might injure people who handle the debris.

To avoid the risks connected with sharps, it is important to use and dispose of sharp objects in a safe and proper manner. Using personal protective equipment, such as gloves, when handling sharps can also lower the risk of injury and pathogen exposure.

7.4 Occurrence of Sharps Injuries

Injuries can happen at any time during its usage, disassembly, or disposal. Injuries caused by solid sharps:

- During usage, specimen processing, collision with sharps, needle insertion or withdrawal, and manipulation of sharpness in the patient.
- After use, prior to disposal, use of the safety function, recapping needles during clean-up, or sharps left in an odd position.
- During the disposal process, poor disposal, or after disposal.
- Equipment design, process type, working conditions, recapping, employee experience, and disposal are few variables influencing these occurrences.

7.5 Prevention of Injuries Caused by Sharps

Sharps injury prevention is regarded as part of any person's normal practices. Sharps users must be aware of how to safeguard themselves and anyone else who can be exposed to the instrument [4]. The most efficient approach to protecting someone is to prevent harm.

- Needles recapping should be avoided, and immediate disposal of sharps in a designated puncture-resistant container should be done.
- Report the occurrence and exposure to higher authorities as soon as possible.
- Look for strategies to limit or eliminate the sharps use.
- Utilizing devices to grab any sharps while using or disposing of them can help prevent injuries.
- If required, personal protective equipment should be used as the last line of protection.
- Creating safe practices for personnel to follow when recapping is judged necessary. Never direct an uncovered hand at an exposed needle tip. To recap, place the cap on a flat surface and spoon it onto the tip of a syringe in one hand. Using your free hand, keep a safe space between the sheath and the exposed needle.
- When using glass slides, needles, glass stir rods, pipettes, blades, scalpels capillary tubes, or any glassware, always use caution.

7.6 Handling of Sharps

- Needles and syringes, as well as other sharp tools, should be used only when there are no other options.
- When feasible, plastic ware should be used instead of glasses.
- When feasible, blunt-ended needles, use needleless systems, safety scalpels, or unbroken glassware.
- Sharps management and disposal procedures must be incorporated into your lab's documented Standard Operating Procedures.
- Needles must not be sheared, bent, recapped, broken, or detached from disposable syringes before discarding.
- Shattered glass should not be picked with hands, instead mechanical tools like a tongs, brush and dustpan, or forceps should be used for clearing.
- Re-capping of the needle can be avoided by carrying it in a tray or protected container. If re-capping the needle is the only option, then use an automatic re-capping mechanical device, or the one-handed scoop method, where the cap is

placed on the counter top and the needle is scooped, keeping the hand free and out of the way.

- If a needle disengages from a syringe, then sprays or aerosols may be produced. To avoid exposure to sprays and aerosols, use needle-locking syringes.
- Disposable scalpels that require you to lay your fingers adjacent to the blade and apply force to break off the shield are not recommended.
- Scalpel should not be sawed or undue force applied to it, which causes the blade to shatter and a flying debris risk as well as a sharps exposure risk.
- When executing an activity that requires a sharp instrument to be handed between workers, employ zones and communication to prevent both personnel from gripping the object at the same moment. Assisting workers should never cross zones and keep their hands as far away from the sharps region as possible. Communication is essential between the sharps handler and the person aiding. This prevents the assisting workers from simultaneously gripping the sharps device.

7.7 Disposal of Sharps

It is vital to have a dependable method for disposing of used needles and sharps in order to avoid damage. Bins for waste should be conveniently accessible. Needles should be kept in puncture-resistant containers with a large opening. Consider if you really need to use sharp instruments like needles and glass pipettes. Skin penetration by a biologically contaminated sharp object can result in the spread of germs and viruses, which can lead to severe diseases [5–7]. Some of the important instructions for disposal of sharps are as follows:

- All sharps must be collected and disposed of in a puncture-proof sharps container. To avoid walking about with infected sharps, keep an adequate sharps container close to the place of usage.
- After using disposable sharps, immediately place them in a sharps container. Before disposing of the device, do not recap needles. If there are no other options, then transfer the

needle and syringe to a leak-proof secondary container with a tight top.

- When cleaning and reprocessing reusable sharps, use cleaning instruments with handles, such as a brush or sponge.
- Sharps should be able to fall freely into the container. If materials do not fall into the container, then it is either too full or too small.
- Do not use the sharps container for gauze, gloves, or other objects that take up space and make it difficult for the sharps to fall easily into the container.
- Never jiggle sharps containers to make extra space. Shaking produces aerosols and can cause objects to spill from containers.
- Sharps should not be pushed into a container, nor should anything be taken out of the container with your hand.
- Sharps containers should not be overfilled. When a container is 50%–60% filled, it is called full. Never allow syringes, needles, or other materials to protrude from the sharps container, which should be firmly closed.
- Place shattered glass in the appropriate containers.
- Keep sharps containers close to workstations for easy and safe disposal.
- When using biohazardous agents with sharps (including plant, animal, or human biohazards), autoclave the sharps container before ultimate disposal.
- Custodial employees and other non-laboratory personnel should never "pick up" sharps for disposal. Sharps disposal is the responsibility of the principal investigators or their designees.

7.8 Reduction of Sharps

Sharps-reduction policies should be implemented in all laboratories. The first step in preventing needle stick and sharps injuries is to implement a sharps reduction policy. Identify any sharps used in procedures and examine whether there is an option to perform the treatment without using an instrument sharp enough to pierce skin. Sharps reduction can be achieved by substituting a standard needle used to scatter clumped cells with a blunt-end or tip needle [8].

7.9 Knowledge About the Environment During Disposal of Sharps

Sharps disposal is critical not just for preventing injuries but also for safeguarding the environment and public health. When sharps are not properly disposed of, they can pose a number of environmental hazards. Here is some crucial environmental information concerning sharps disposal:

a) Sharps in Landfills: Sharps, particularly needles, can represent a major risk to waste management employees and the environment if they are disposed of with normal trash and wind up in landfills. Needles can puncture garbage bags, potentially exposing employees to injury and blood borne diseases.
b) Water Source Contamination: Sharps that are improperly disposed of in bodies of water or flushed down toilets can pollute water supplies. This can result in the spread of infectious illnesses and the contamination of the environment.
c) Hazardous Materials: Some sharps may contain dangerous chemicals or drugs. When these compounds are improperly disposed of, they can leak into the land and water, possibly hurting plants, animals, and aquatic life.
d) Impact on Wildlife: Improperly dumped sharps can endanger wildlife, particularly marine creatures. Animals may be injured or exposed to diseases if they come into contact with discarded sharps on beaches or in rivers.
e) Needlestick litter: It is defined as abandoned needles found in public areas such as parks, playgrounds, and sidewalks. This poses a major risk to the general public, particularly youngsters and unwary individuals who may tread on or contact the sharps by mistake.
f) Air Pollution: When sharps are cremated without sufficient measures, dangerous chemicals and germs are released into the air, compromising air quality.
g) Microplastics: Some sharps may have plastic components, such as syringes with plastic barrels. When these products are inappropriately disposed of, they can add to the rising problem of microplastic contamination in the environment.

To reduce the environmental impact of sharps disposal, it is critical to follow the following guidelines:

- Sharps containers should be utilized for disposal, especially at educational and healthcare institutions, and other areas where sharps are often handled.
- Sharps should not be disposed of in ordinary garbage, recycling containers, or down the toilet.
- If you come across a discarded sharp in a public place, do not pick it up with your hands; instead, report it to the relevant authorities for safe disposal.
- Educate the public on the necessity of safe sharps disposal and the dangers of inappropriate disposal.
- Encourage the use of sharps with safety features to help reduce injuries and encourage proper disposal.
- We can protect both public health and the natural world by disposing of sharps in an environmentally conscious manner.

7.10 First Aid for Accidents Caused Due to Sharps

Sharps-related injuries, such as those caused by needles, knives, or shattered glass, can be fatal owing to the danger of infection and bleeding. If you are injured by a sharp item, then perform these first-aid steps:

- Protect yourself from any potential exposure to blood or body fluids before aiding the wounded individual. If disposable gloves are available, then wear them.
- If the person is bleeding, then use a clean cloth or sterile gauze pad to apply direct pressure to the lesion. Avoid removing the item if it is still entrenched in the wound. Apply pressure around the item instead to stop the bleeding.
- Elevate the damaged location above the level of the heart if feasible. This can assist in reducing blood flow to the site and slowing bleeding.
- After the bleeding has stopped, carefully clean the area surrounding the cut with mild soap and water. Scrubbing the wound may cause more injury.

- If available, cover the wound with a sterile dressing or a clean towel. If the item is still lodged in the wound, then avoid directly covering it.
- It is critical to get expert medical attention for any injury produced by a sharp item, particularly if:
  - The wound is huge, deep, or gaping.
  - The item has become entrenched in the wound.
  - Applying pressure does not halt the bleeding.
  - The individual is in excruciating pain or has trouble moving the affected region.
  - The injury occurs on the face, neck, or in close proximity to important organs.
- Squeezing the wound to remove blood or fluid may raise the risk of infection. Similarly, avoid sucking on the wound because this might introduce deadly bacteria.
- Only use tourniquets if you are in a life-threatening emergency and are unable to stop the bleeding by other methods. If applied incorrectly, then tourniquets can cause significant injury.
- Reassure and soothe the injured individual while waiting for medical assistance.
- Apply milk to the wound and carefully clean it with soap and water. Then, wrap it in a bandage.
- Any inadvertent injection of a solvent, especially a dangerous one like dichloromethane, should be regarded extremely seriously and action done promptly. If the color of the puncture hole and surrounding region changes to pink or purple in a short amount of time, this is most likely a symptom of necrosis (death of bodily tissue), and medical evaluation and treatment must be sought immediately; emergency surgery may be necessary.
- Inform your supervisor as soon as possible about the injury and seek medical advices.

Remember that first aid is not a replacement for professional medical care. To ensure correct diagnosis, treatment, and prevention of consequences, always seek quick medical assistance for serious injuries caused by sharps.

References

1. Trim, J.C. and Elliott, T.S., A review of sharps injuries and preventative strategies. *J. Hosp. Infect.*, 53, 4, 237–242, 2003.
2. Spruce, L., Back to basics: sharps safety. *AORN J.*, 104, 1, 30–36, 2016.
3. Pugliese, G., Germanson, T.P., Bartley, J., Luca, J., Lamerato, L., Cox, J., Jagger, J., Evaluating sharps safety devices meeting OSHA's intent. *Infect. Control Hosp. Epidemiol.*, 22, 7, 456–458, 2001.
4. Ford, D.A., Implementing AORN recommended practices for sharps safety. *AORN J.*, 99, 1, 106–120, 2014.
5. Link, T., Guidelines in Practice: Sharps Safety. *AORN J.*, 112, 5, 478–489, 2020.
6. Perry, J., Robinson, E.S., Jagger, J., Nursing2004 needle-stick and sharps-safety survey: getting to the point about preventable injuries. *Nursing2023*, 34, 4, 43–47, 2004.
7. Phillips, E.K., Conaway, M., Parker, G., Perry, J., Jagger, J., Issues in understanding the impact of the Needlestick Safety and Prevention Act on hospital sharps injuries. *Infect. Control Hosp. Epidemiol.*, 34, 9, 935–939, 2013.
8. Markkanen, P., Galligan, C., Laramie, A., Fisher, J., Sama, S., Quinn, M., Understanding sharps injuries in home healthcare: The Safe Home Care qualitative methods study to identify pathways for injury prevention. *BMC Public Health*, 15, 1–2, 2015.

8

Safety Equipment in Academic Institution Laboratories and Workshops

Abstract

Laboratory and workshop safety equipment is of paramount importance to ensure the well-being of researchers, students, and laboratory personnel. This chapter explores the various types of lab safety equipment designed to mitigate potential hazards and promote a secure working environment. The discussion encompasses personal protective equipment, emergency equipment, fire safety gear, and specialized equipment for handling hazardous materials. The final section covers the importance of record keeping and documentation for maintaining a safe laboratory environment, to ensure regulatory compliance and support incident investigations.

***Keywords*:** Lab safety, workshop safety, safety equipment, maintenance

8.1 Background

The scientific research and experimentation world is fascinating and dynamic, full of exploration, innovation, and discovery. However, working in a laboratory comes with its fair share of risks, as hazardous chemicals, high temperatures, and dangerous equipment can pose serious threats to the health and safety of researchers and staff. Every laboratory must have the right safety gear and equipment to mitigate these risks and ensure a safe working environment.

From personal protective gear to emergency response tools, we will provide an overview of the most essential items to maintain a secure and efficient workspace.

Abhishek Barua, Kanchan Kumari, Sumit Saha, Siddharth Jeet and Swastik Pradhan. Institutional and Industrial Safety Engineering Practices, (111–178) © 2025 Scrivener Publishing LLC

8.2 Introduction

To work in a laboratory, you must first make it safe. You are in great danger of terrible mishaps if you do not provide lab safety equipment. Laboratories are inherently risky environments. There are many dangerous gadgets, such as Bunsen burners that operate with gas-operated open flames, corrosive acids and bases, spring-loaded equipment that can impact people, sharp gear that can slash your hand and infect you, and the list goes on and on. Because all of these factors (and more) make a laboratory a dangerous environment, investing in lab safety equipment early on is your best chance. Laboratory safety equipment is critical for safeguarding researchers, students, and laboratory personnel from various risks in the laboratory. There are various types of safety equipment designed to address different risks and ensure a safe working environment. Here are some common lab safety equipment items [1].

8.3 Importance of Lab and Workshop Safety Equipment

The importance of lab safety cannot be overstated. Laboratories are places where experiments and research are conducted, often involving hazardous materials, high-energy equipment, and potentially dangerous processes [2]. Prioritizing safety in these environments helps to:

- Protection of researchers and staff: Ensuring the well-being of those working in the laboratory is the primary concern of lab safety. Proper safety measures help prevent accidents, injuries, and long-term health issues arising from exposure to harmful substances or dangerous situations.
- Preservation of research integrity: Accidents and mishaps in the laboratory can compromise research outcomes, wasting valuable time, resources, and effort. A safe laboratory environment contributes to the research findings' accuracy, reliability, and validity.
- Complied with ethical and legal requirements: Many countries have regulations and guidelines in place to ensure the safety of laboratory workers and the environment. Adhering to these rules helps maintain a safe workspace, avoids legal consequences, and maintains the institution's reputation.

- Minimization of environmental impact: Proper handling and disposal of hazardous materials and waste are essential to reducing the environmental impact of laboratory activities. Lab safety measures contribute to a more sustainable and eco-friendly research environment.

8.4 Purpose of Lab and Workshop Safety Equipment

In the laboratory, safety equipment fulfills numerous critical functions, including:

- Accident preventative measures: safety shields, fume hoods, and chemical storage cabinets reduce the risk of accidents by storing hazardous materials and minimizing possible risks like fires or explosions.
- Accident protection: Personal protective equipment (PPE) protects researchers and personnel from direct contact with hazardous chemicals, lowering the risk of harm and contamination. Gloves, safety goggles, coats for the lab, and respirators are a few examples.
- Emergency responses: In the event of an accident or emergency, safety equipment such as eyewash stations, safety showers, firefighting equipment, and spill containment kits enable for a rapid and effective response, minimizing harm and damage.
- Keeping a clean and organized workspace: A well-organized laboratory decreases the likelihood of accidents and contamination. Storage containers, labeling, and trash disposal methods all contribute to the lab's order and cleanliness.

Overall, lab safety equipment is critical to establishing a safe, efficient, and responsible research environment that protects both staff and scientific activity.

8.5 Different Types of Safety Equipment

Every laboratory should have several pieces of laboratory safety equipment to minimize the risk of accidents and ensure the safety of laboratory workers [3–5]. Here are some of the most important types of lab safety equipment:

8.5.1 Personal Protective Equipment (PPE)

Personal Protective Equipment (PPE) is crucial for laboratory safety, as it helps to protect workers from potential hazards that could cause injury or illness. Some of the essential PPE items that every laboratory should have include the following:

a) Safety goggles: Safety goggles are an essential piece of PPE used in laboratories to protect the eyes from potential hazards (see Figure 8.1). They provide a barrier between the eyes and various materials or substances that could cause eye injuries, such as chemicals, flying debris, or biological agents. Here are some important points to consider regarding safety goggles in laboratories:

- Eye Protection: Safety goggles are specifically designed to provide complete coverage around the eyes, including the sides, offering protection against splashes, spills, and airborne particles.
- Chemical Splash Protection: In chemistry labs, safety goggles protect against accidental chemical splashes that can cause irritation, burns, or long-term damage to the eyes.

Figure 8.1 Safety goggles.

- Impact Protection: Safety goggles are made from impact-resistant materials to shield the eyes from objects that may be propelled or fall, such as broken glassware or metal shavings.
- Compatibility With Other PPE: Safety goggles are often designed to be worn in combination with other PPE, such as lab coats and gloves, for comprehensive protection.
- Types of Safety Goggles: There are different types of safety goggles available, depending on the specific hazards in the laboratory. For instance, there are safety goggles with indirect vents to prevent chemical splashes from entering, and there are also sealed safety goggles for protection against biohazards or infectious materials.
- Fit and Comfort: Properly fitted safety goggles offer a snug and comfortable fit, reducing the likelihood of slippage or discomfort during prolonged use.
- Fog Resistance: Some safety goggles come with anti-fog coatings or ventilation features to prevent fogging, which can impede vision and compromise safety.
- Regular Inspection: Safety goggles should be inspected regularly for signs of damage, such as scratches or cracks, and replaced if compromised to maintain their effectiveness.
- User Training: Laboratory personnel should receive training on how to properly wear, adjust, and care for safety goggles to ensure that they provide adequate protection.
- Use with Prescription Eyewear: Safety goggles can be designed to accommodate prescription glasses or can be equipped with prescription lenses to ensure proper vision for individuals who need corrective eyewear.
- Restricted Areas: In some laboratories, specific areas or tasks may require the use of full-face protection or specialized goggles to address unique hazards.

Safety goggles are an integral part of laboratory safety, and wearing them consistently when handling hazardous materials or conducting experiments is crucial to prevent eye injuries and promote a safe work environment.

b) Lab coats: Lab coats are an important piece of PPE used in laboratories to provide a protective barrier between laboratory workers and potential hazards. They serve multiple purposes and are a fundamental component of lab safety. They are a fundamental part of lab safety protocols and should be worn consistently when working with hazardous materials or conducting experiments in the laboratory. Here are the key features and benefits of lab coats:

- Clothing Protection: Lab coats are designed to cover the front of the body, from the neck down to the knees or lower. They protect the wearer's regular clothing from spills, splashes, and potential contamination with hazardous substances, including chemicals, biological agents, and radioactive materials.
- Material: Lab coats are typically made from durable and non-absorbent materials, such as cotton, polyester, or a combination of fabrics. This helps to prevent the penetration of hazardous substances through the coat. The choice of lab coat material will depend on the specific needs and hazards of the laboratory environment. For general-purpose lab work, a blend of polyester and cotton or pure cotton lab coats may be suitable. In more hazardous environments, such as those involving chemicals or flames, specialized lab coat materials like Nomex may be necessary to provide adequate protection (see Table 8.1).
- Full-Length Coverage: The full-length design of lab coats provides comprehensive protection for the upper and lower body, reducing the risk of skin exposure to hazardous materials.
- Easy Removal: Lab coats are often designed to be easily removed in case of chemical spills or accidents to prevent further exposure.
- Front Closure: Most lab coats have a front button or snap closure, allowing for quick and secure fastening and removal.
- Compatibility With Other PPE: Lab coats are designed to be worn over regular clothing and can be easily combined with other PPE, such as safety goggles, gloves, and closed-toe shoes.
- Identification: In some laboratory settings, lab coats may have a specific color or feature identifiers, such as embroidered names or badges, to help identify personnel or specific roles within the lab.
- Lab-Specific Design: Lab coats are tailored to meet the unique needs of laboratory environments, with features that enhance functionality and safety.

Table 8.1 Different lab coat materials.

Material	Properties
Cotton	Cotton lab coats are comfortable, breathable, and lightweight. They provide moderate protection against spills and splashes but may not offer adequate chemical resistance.
Polyester	Polyester lab coats are resistant to stretching and shrinking, and they hold their shape well over time. They also offer better chemical resistance compared to cotton.
Polyester/Cotton Blends	Combining polyester with cotton results in lab coats that are more durable, wrinkle-resistant, and colorfast compared to pure cotton options.
Polycotton	A blend of polyester and cotton, polycotton lab coats combine the durability and wrinkle resistance of polyester with the comfort of cotton.
Polypropylene	Polypropylene lab coats are commonly used as disposable options, offering basic protection against spills and contaminants.
Tyvek	Tyvek lab coats are lightweight, durable, and provide a barrier against particles and some chemicals. They are often used as disposable lab coats.
Nomex	Nomex lab coats are flame-resistant and are often used in laboratories where there is a risk of exposure to flames or high heat.

- Regular Cleaning: Proper maintenance and regular cleaning of lab coats are crucial to prevent cross-contamination and maintain their effectiveness as protective clothing.
- Proper Storage: When not in use, lab coats should be stored in designated areas, separate from personal clothing or areas where hazardous materials are present.
- Restricted Usage: In some laboratories, certain areas or tasks may require the use of specialized lab coats or disposable lab coats to address specific hazards.
- Color code: In some laboratory settings, lab coats may be assigned different colors to signify specific roles, responsibi-

lities, or safety protocols (see Table 8.2 and Figure 8.2). The use of different colored lab coats can help with easy identification and improve communication within the laboratory environment. However, the specific color-coding system may vary between institutions and laboratories. It is

Table 8.2 Different colors of lab coat according to the usage.

Usage type	Color
General Lab Personnel	White lab coats are often worn by general lab personnel, including researchers, scientists, and students. White is a common color for standard lab coats and signifies general laboratory work.
Visitors or Non-Lab Personnel	In certain laboratories, visitors, guests, or non-lab personnel may be required to wear lab coats of a specific color (e.g., light blue or gray) to indicate that they are not directly involved in laboratory work.
Supervisors or Principal Investigators	In some institutions, supervisors or principal investigators may wear a different color lab coat, such as navy blue or black, to distinguish their leadership roles.
Technicians or Support Staff	Lab coats in a different color, such as green or gray, might be worn by technicians or support staff to differentiate their roles from researchers or scientists.
Contamination Control	In certain labs, disposable lab coats in distinct colors (e.g., red or yellow) might be used for specific tasks involving potentially hazardous or contaminated materials. These disposable coats are disposed of after each use to prevent cross-contamination.
Specialized Roles	In specific research areas or high-security laboratories, different colors may be used to designate personnel with specialized training, such as biological safety specialists (e.g., yellow lab coats) or radiation safety officers (e.g., orange lab coats).

Figure 8.2 Different colors of lab coat.

essential to note that color-coding systems for lab coats can differ between institutions, and some laboratories may not use color-coded lab coats at all. This will ensure that lab personnel, visitors, and other individuals can easily identify roles and responsibilities while maintaining a safe and efficient working environment.

c) Gloves: In both laboratories and workshops, various types of gloves are used to provide protection against specific hazards and ensure the safety of workers (see Table 8.3). The choice of gloves depends on the nature of the work being performed and the potential risks involved.

Table 8.3 Different types of gloves.

Material	Figure	Description
Nitrile Gloves	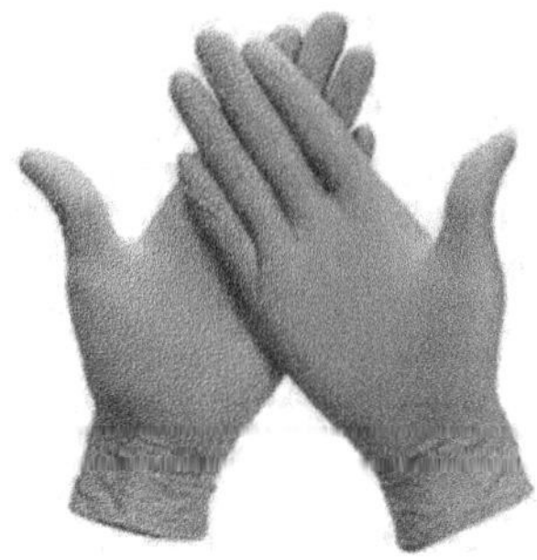	Nitrile gloves offer excellent resistance to a wide range of chemicals, making them suitable for handling various hazardous substances in both laboratories and workshops. Nitrile gloves are latex-free, reducing the risk of allergic reactions for individuals sensitive to latex.
Neoprene Gloves	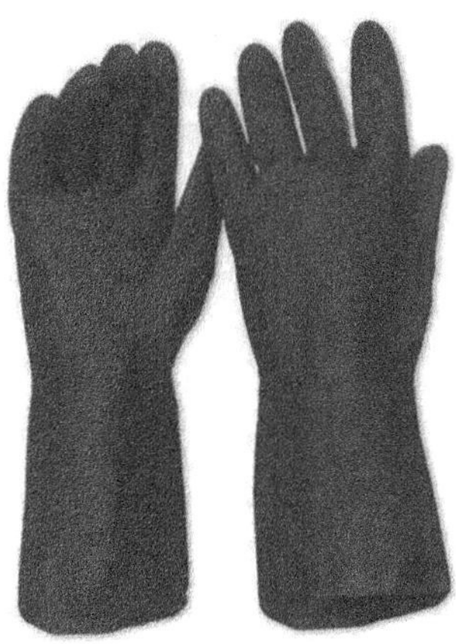	Neoprene gloves offer good resistance to chemicals, solvents, oils, and greases, making them suitable for tasks involving these substances.

(*Continued*)

Table 8.3 Different types of gloves. (*Continued*)

Material	Figure	Description
Latex Gloves	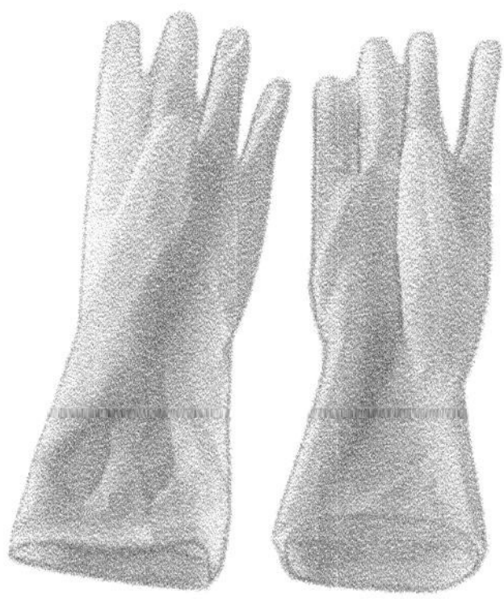	Latex gloves provide a snug fit and offer good dexterity, making them ideal for tasks that require precision and fine motor skills. Latex gloves may cause allergic reactions in some individuals, so alternatives like nitrile or neoprene gloves are preferred for those with latex sensitivity.
Vinyl Gloves		Vinyl gloves are cost-effective and commonly used for general-purpose tasks in workshops and laboratories where chemical resistance is not a primary concern. They provide limited chemical resistance compared to nitrile or neoprene gloves.

(*Continued*)

Table 8.3 Different types of gloves. (*Continued*)

Material	Figure	Description
Polyethylene Gloves	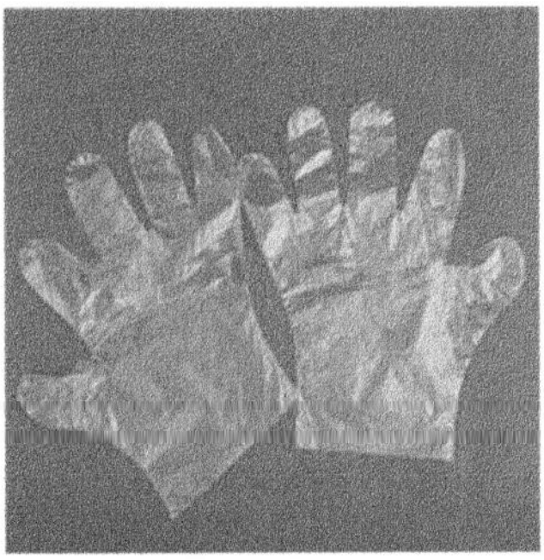	Polyethylene gloves are lightweight and disposable, often used in workshops and labs for tasks that require frequent glove changes to prevent contamination.
Butyl Rubber Gloves	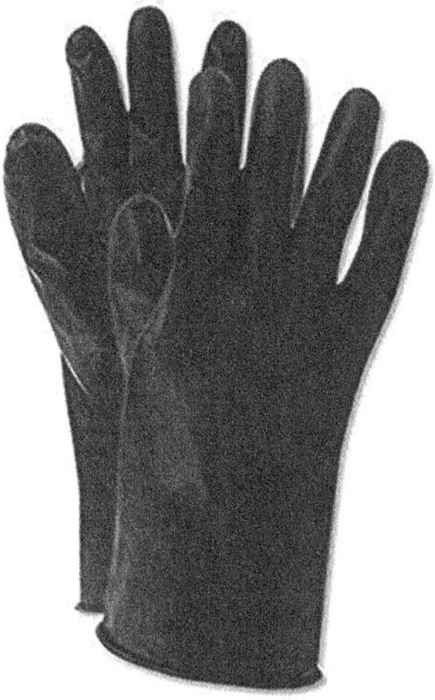	Butyl rubber gloves provide exceptional resistance to corrosive chemicals, acids, and gases, making them suitable for highly hazardous tasks.
Leather Gloves		Leather gloves offer protection against abrasions, cuts, and punctures, making them suitable for heavy-duty tasks in workshops.

(*Continued*)

Table 8.3 Different types of gloves. (*Continued*)

Material	Figure	Description
Heat-Resistant Gloves	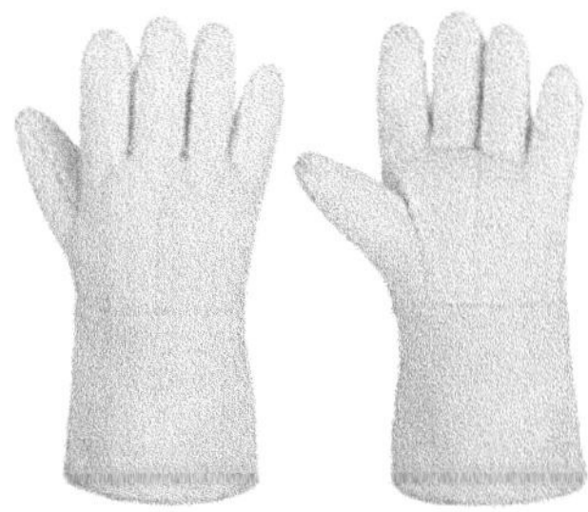	Heat-resistant gloves are used in workshops where there is a risk of exposure to high temperatures, such as welding or handling hot objects.
Cut-Resistant Gloves	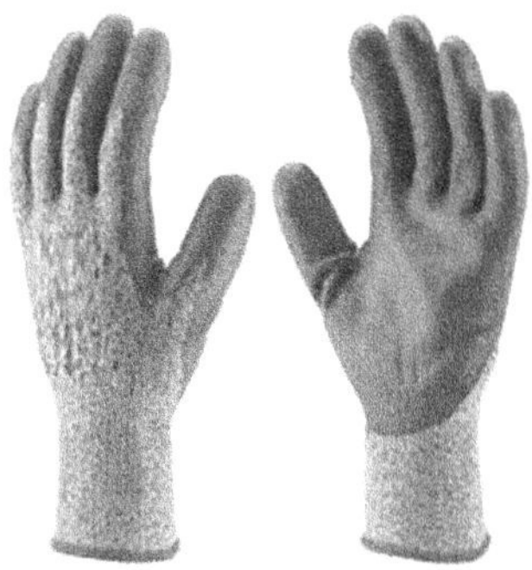	Cut-resistant gloves are designed to protect against sharp objects, such as glass or metal, and are commonly used in workshops for tasks involving cutting or handling sharp tools.

It is essential to choose the appropriate type of glove based on the specific hazards and tasks involved. Proper glove selection, along with training on their correct use and disposal, is crucial for maintaining a safe working environment in both laboratories and workshops.

d) Face masks and respirators: Face masks and respirators are crucial pieces of PPE used in both laboratories and workshops to protect workers from respiratory hazards. These respiratory protection devices help filter out harmful airborne particles, dust, and contaminants that could pose health risks. The choice of face masks and respirators depends on the level of protection required and the nature of the hazards present. Here are different types commonly used (see Table 8.4).

It is essential to select the appropriate respiratory protection based on the specific hazards present in the laboratory or workshop. Training in the correct use, fitting, and maintenance of these devices is essential to ensure their effectiveness and to promote a safe working environment. Additionally, regular inspection, proper storage, and adherence to

Table 8.4 Different types of face mask.

Mask type	Figure	Description
N95 Respirators	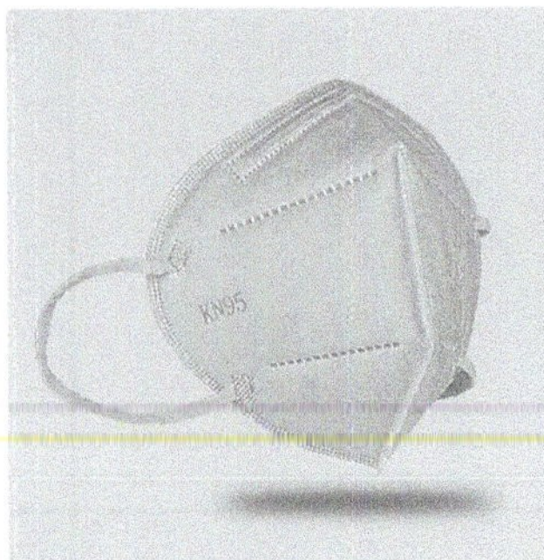	N95 respirators are tight-fitting respirators that filter at least 95% of airborne particles, including small aerosolized particles. They protect both the wearer and others from respiratory hazards. N95 respirators are essential in environments where airborne particles, such as dust, fumes, or bioaerosols, are present. They are commonly used in workshops where tasks generate particulates or in labs handling hazardous materials.
Surgical Masks	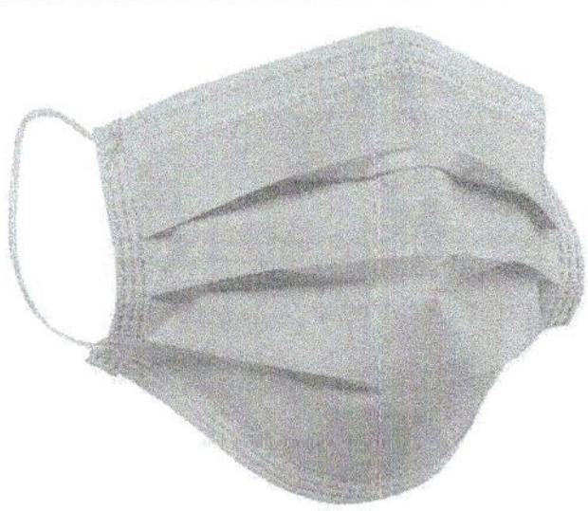	Surgical masks are loose-fitting disposable masks that primarily protect others from respiratory droplets produced by the wearer. They are commonly used in healthcare settings but might be used in workshops for certain tasks that do not involve hazardous particles.

(Continued)

Table 8.4 Different types of face mask. (*Continued*)

Mask type	Figure	Description
P100 Respirators		Exceptional Filtration: P100 respirators offer the highest level of filtration efficiency (99.97%) and also provide protection against oil-based particles. P100 respirators are used in environments with oil-based aerosols or when dealing with particularly hazardous airborne particles.
N99 and N100 Respirators	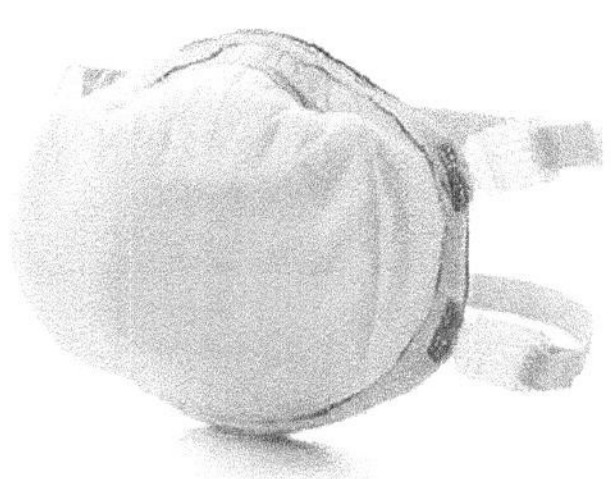	N99 and N100 respirators offer even higher filtration efficiency (99% and 99.97%, respectively) than N95 respirators, providing enhanced protection against fine particulates. These respirators are used in environments with more significant airborne particulate hazards, such as highly contaminated workshops or labs.

(*Continued*)

Table 8.4 Different types of face mask. (*Continued*)

Mask type	Figure	Description
Half-Face and Full-Face Respirators		Half-face respirators cover the nose and mouth, whereas full-face respirators also protect the eyes. Half-face respirators are commonly used in workshops where respiratory protection is needed, whereas full-face respirators are used in situations where eye protection is also required.
Powered Air-Purifying Respirators (PAPRs)		PAPRs are battery-powered respirators that use a fan to draw air through filters for breathing. PAPRs are used in situations where a higher level of comfort and protection is required for extended periods, such as during lengthy workshops or tasks with intense respiratory hazards.

manufacturer guidelines are crucial for maintaining the integrity of face masks and respirators [5–8].

e) Closed-toe shoes: Closed-toe shoes are a fundamental part of PPE in both laboratories and workshops. They provide protection for the feet from potential hazards, including falling objects, spills, sharp materials, and chemicals. Here are some reasons why closed-toe shoes are essential in lab and workshop settings:

- Protection From Falling Objects: In both laboratories and workshops, there is a risk of objects or tools accidentally falling on the feet. Closed-toe shoes offer a protective barrier against such incidents, reducing the risk of injuries.
- Spill Protection: Closed-toe shoes prevent liquids, chemicals, and other hazardous substances from coming into direct contact with the feet, providing an additional layer of protection against spills.
- Chemical Resistance: Some closed-toe shoes are made from materials that offer chemical resistance, which is particularly important in laboratories where researchers and technicians may handle potentially corrosive or hazardous chemicals.
- Sharp Object Protection: Workshops often involve handling tools and materials that could cause injuries if stepped on. Closed-toe shoes with sturdy soles help protect against such sharp objects.
- Puncture Resistance: Certain closed-toe shoes are designed to be puncture-resistant, providing an added level of safety in workshops where sharp objects may be present.
- Slips and Trips Prevention: Closed-toe shoes with slip-resistant soles can help reduce the risk of accidents caused by slips and trips, which can occur in both laboratories and workshops due to spills or uneven surfaces.
- Compliance with Safety Regulations: Many workplace safety regulations require the use of closed-toe shoes in laboratories and workshops as a standard safety measure.

When selecting closed-toe shoes for lab and workshop use, it is essential to consider the specific hazards and conditions of the environment (see Table 8.5). The shoes should be comfortable, properly fitting, and appropriate for the tasks performed. Employees should be educated on the importance of wearing closed-toe shoes and the role they play in maintaining a safe working environment. Regular inspection and replacement of worn-out or damaged shoes are also important for ensuring continued protection.

Remember that the specific type of closed-toe shoes needed will depend on the particular hazards present in the laboratory or workshop environment. Additionally, it is crucial to ensure that the shoes fit properly, are comfortable for extended wear, and comply with any relevant safety standards or regulations. Regular inspection and maintenance of footwear are also essential to ensure their continued effectiveness in protecting the feet from potential dangers.

Table 8.5 Different types of shoes.

Mask type	Figure	Description
Steel Toe Work Boots		Steel toe work boots have a reinforced toe cap made of steel, providing protection against falling objects or heavy materials in workshops and construction sites. Some steel toe work boots also provide electrical hazard protection to safeguard against electrical currents.
Laboratory Safety Shoes		Laboratory safety shoes often have slip-resistant soles to provide better traction on smooth and wet surfaces, reducing the risk of slips and falls. Some laboratory shoes are made from chemical-resistant materials to protect against accidental spills of hazardous substances. The closed-heel and toe design of these shoes offers full coverage and protection for the foot.

(*Continued*)

Table 8.5 Different types of shoes. (*Continued*)

Mask type	Figure	Description
Composite Toe Work Shoes		Composite toe work shoes have a toecap made of non-metallic materials (such as carbon fiber or Kevlar) that provide impact protection without adding the weight of steel. Composite toe shoes do not conduct electricity, making them a safer option in electrical work environments.
Chemical-Resistant Boots		For environments with high chemical exposure, chemical-resistant boots made of materials like PVC or neoprene are suitable for protecting the feet and lower legs. In environments with potential exposure to liquids, these boots are waterproof and beneficial for keeping feet dry and protected.

(*Continued*)

Table 8.5 Different types of shoes. (*Continued*)

Mask type	Figure	Description
Anti-Static Shoes		Anti-Static shoes are designed to discharge static electricity safely, protecting sensitive electronic equipment in laboratories or workshops with electrostatic dissipative sensitive components.
Oil-Resistant Sole shoes		In workshops with oily or greasy surfaces, shoes with oil-resistant soles provide better traction and reduce the risk of slips.
Slip-On Work Shoes		Slip-on work shoes are convenient and quick to put on and remove, making them suitable for environments where frequent entry and exit are required.

f) Hearing protection: Hearing protection is essential in both laboratories and workshops where noise levels can exceed safe limits. Prolonged exposure to loud noises can lead to permanent hearing damage, so it is crucial to provide appropriate hearing protection for workers. When selecting hearing protection, it is essential to consider the noise levels in the lab or workshop and choose the appropriate level of protection based on the decibel rating. Proper fit and comfort are also critical to ensure workers will wear the hearing protection consistently (see Table 8.6). Here are different types of hearing protection commonly used in labs and workshops.

Training workers on the correct use, maintenance, and storage of hearing protection is crucial to maximize its effectiveness. Regular inspections and replacement of damaged or worn-out hearing protection are

Table 8.6 Different types of ear protection.

Mask type	Figure	Description
Earplugs		These are affordable and made from foam that expands to fit the ear canal, providing a snug seal for single use. Reusable silicone earplugs are made from soft, washable silicone and can be reused multiple times.
Earmuffs		These have cushioned ear cups that fit over the ears and an adjustable headband for a secure fit. Behind-the-neck earmuffs are similar to over-the-head earmuffs, but the headband fits behind the neck for comfort when wearing other headgear, such as hard hats. Electronic earmuffs provide active noise reduction by using built-in microphones to pick up ambient sounds and play them back through the ear cups, canceling out harmful noises while allowing communication.

(*Continued*)

Table 8.6 Different types of ear protection. (*Continued*)

Mask type	Figure	Description
Custom-Molded Earplugs	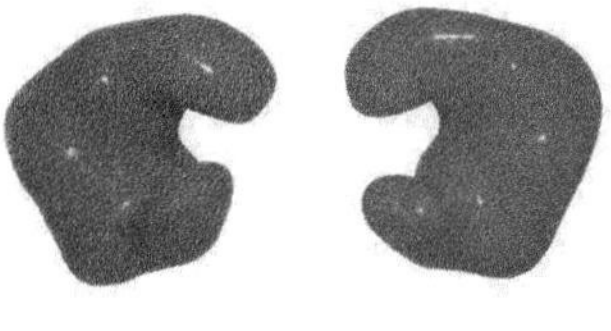	Custom-molded earplugs are individually made to fit the unique shape of the wearer's ear, providing a comfortable and secure fit for long-term use.
Canal Caps	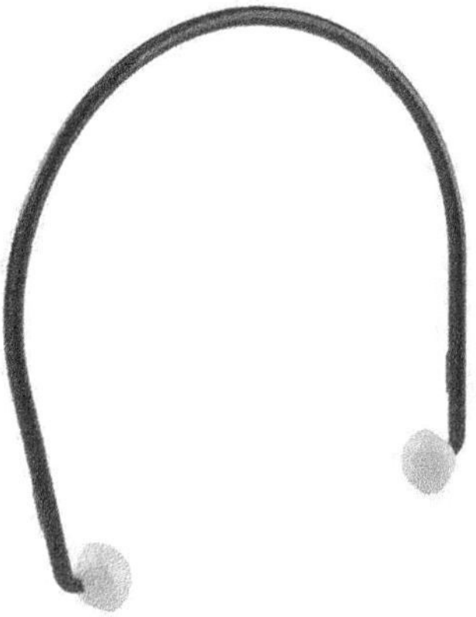	Canal caps combine the benefits of earplugs and earmuffs, offering a lightweight design that fits partially in the ear canal and partially over the ear.
Noise-Canceling Headphones		Noise-canceling headphones use active technology to reduce ambient noise, providing hearing protection in environments with lower noise levels.

also necessary to ensure continued protection against hazardous noise levels.

8.5.2 Fire Safety Equipment

Fire safety equipment is crucial in any laboratory, as fires can be caused by various hazards, such as electrical equipment, flammable chemicals, or heating sources. The following are some of the essential fire safety equipment that every laboratory should have:

a) Fire extinguishers: For laboratories and workshops, it is essential to have a combination of fire extinguishers that can handle the potential types

Table 8.7 Different classes of fire and suitable fire extinguisher.

Fire type	Description	Suitable fire extinguisher
Class A	These fires involve common combustible materials such as wood, paper, cloth, and plastics.	Water (pressurized water), foam [aqueous film-forming foam (AFFF)], or dry chemical (Multi-purpose) extinguishers are effective for class A fires.
Class B	These fires involve flammable liquids or gases, such as gasoline, oil, grease, solvents, or propane.	Carbon dioxide (CO_2) or dry chemical extinguishers are appropriate for class B fires.
Class C	These fires involve energized electrical equipment, where the use of water-based extinguishers is unsafe due to the risk of electrical shock.	Carbon dioxide (CO_2) or dry chemical extinguishers are effective for class C fires.
Class D	These fires involve combustible metals, such as magnesium, titanium, or potassium. Class D fires are less common in laboratories and workshops and usually require specialized extinguishing agents.	Class D fires require specific extinguishing agents designed for the particular metal involved. Some examples include specialized powder or sand for specific metals.

of fires that may occur. There are four primary classes of fires and specialized fire extinguishers are required for specific hazards unique to the laboratory or workshop setting (see Table 8.7).

It is crucial to provide appropriate fire extinguisher training to all personnel so that they can use the extinguishers effectively and safely in the event of a fire. Regular maintenance and inspection of fire extinguishers are also necessary to ensure that they are in good working condition when needed. Additionally, fire evacuation plans and emergency procedures should be established and communicated to all occupants of the laboratory or workshop.

b) Fire blankets: Fire blankets are a valuable safety tool in both laboratories and workshops as they can effectively smother small fires and prevent them from spreading. Fire blankets are made from fire-resistant materials and are designed to be used to extinguish fires involving flammable liquids, clothing, and certain other materials. Here is how fire blankets can be used in labs and workshops:

- Fire on a Person: In the event of a fire on a person's clothing, a fire blanket can be wrapped around them to smother the flames and protect the individual from further injury. It is essential for lab and workshop personnel to be trained in the proper use of fire blankets in such emergency situations.
- Small Fires on Surfaces: Fire blankets can also be used to cover small fires on surfaces, such as flammable liquids that have spilled or ignited. By placing the blanket over the fire source, the oxygen supply is cut off, suffocating the flames and preventing the fire from spreading.
- Flammable Material Storage: In workshops, where flammable materials may be stored, having a fire blanket nearby can provide an additional layer of protection in case of fire emergencies.
- Welding and Hot Work: In workshops where welding or other hot work is performed, fire blankets can be used to shield nearby combustible materials from sparks or splatter.

It is essential to ensure that fire blankets are easily accessible and located in visible and easily reachable locations in both laboratories and workshops. Additionally, training all personnel on the proper use of fire blankets and their limitations is crucial to responding effectively to fire emergencies. Remember that while fire blankets are useful for handling certain types of fires, they have their limitations and may not be effective for large or spreading fires. In case of significant fires or if the fire is out of control, evacuation and the use of appropriate fire extinguishing methods should be prioritized while ensuring personal safety. Having a comprehensive fire safety plan, including fire extinguishers, fire blankets, and evacuation procedures, is crucial for creating a safe working environment in both labs and workshops.

c) **Fire alarm systems:** Implementing a reliable fire alarm system is critical for ensuring the safety of occupants in laboratories and workshops. A well-designed fire alarm system can detect fires early, alert individuals to evacuate the area, and provide necessary information to emergency responders. When setting up a fire alarm system for labs and workshops, consider the following components:

- Smoke Detectors: Install smoke detectors throughout the laboratory and workshop areas. Smoke detectors sense the presence of smoke particles and trigger the fire alarm when smoke levels reach a certain threshold. These detectors are essential for early fire detection, especially in areas where flammable materials are stored or used.
- Heat Detectors: Heat detectors are suitable for environments where smoke detectors may be prone to false alarms due to dust, fumes, or other factors. Heat detectors activate the alarm system when a predetermined temperature threshold is exceeded, indicating the presence of a fire.
- Manual Pull Stations: Install manual pull stations at readily accessible locations throughout the lab and workshop areas. These stations allow individuals to manually activate the fire alarm system in case of an emergency.
- Alarm Notification Devices: Fire alarm systems should be equipped with loud horns, sirens, or strobe lights to provide audible and visual alerts to occupants in the event of a fire. The alarms should be sufficiently loud and visible to ensure that they can be heard and seen throughout the entire area.
- Control Panel: The fire alarm system should have a central control panel that monitors and manages all the connected fire alarm devices. The control panel displays the status of each device and provides necessary information for emergency responders.
- Monitoring System: Consider integrating the fire alarm system with a monitoring service that can automatically alert

emergency services when the alarm is activated. This allows for a quicker response time to potential fire incidents.
- Emergency Communication: If feasible, incorporate emergency communication systems, such as intercoms or emergency phones, to allow for clear communication during fire emergencies.
- Battery Backup: Ensure the fire alarm system has a reliable battery backup to continue functioning during power outages.
- Regular Testing and Maintenance: Conduct regular testing and maintenance of the fire alarm system to ensure it remains fully functional at all times.
- Compliance With Regulations: Ensure that the fire alarm system meets all relevant local and national fire safety regulations and codes.

It is essential to work with qualified fire safety professionals and follow their recommendations when designing and installing a fire alarm system tailored to the specific needs of the laboratory and workshop environment. Regular fire drills and fire safety training for all occupants are also crucial to ensure everyone knows how to respond appropriately in case of a fire emergency.

d) Emergency exits and signage: Emergency exits and signage play a critical role in ensuring the safety of individuals in laboratories and workshops. They help guide people to quickly and safely evacuate the premises during emergencies, such as fires, chemical spills, or other hazardous incidents. Here are important considerations for emergency exits and signage in labs and workshops:

- Emergency Exit Locations: Clearly mark all emergency exit locations with illuminated signs that are visible from different areas within the facility. Make sure the exits are easily accessible and unobstructed at all times.
- Exit Doors: Emergency exit doors should open outward and be equipped with panic bars or push bars for quick and easy egress. They should be kept free from obstructions, and their locks should allow for immediate exit without the need for keys.
- Exit Routes: Designate clear and unambiguous exit routes with visible signs, guiding individuals to the nearest

emergency exits. The paths should be well-lit and free from any potential hazards.

- Exit Signage: Install illuminated exit signs above all emergency exit doors and along the designated exit routes. The signs should use universally recognized symbols to indicate the direction of the exit.
- Exit Sign Lighting: Ensure that the exit signs have backup power sources, such as battery-powered lights or emergency generators, to remain visible during power outages.
- Color Codes: Consider using color codes to differentiate emergency exit signs from other signage in the facility. Green is a commonly used color for emergency exit signs.
- Floor Plans and Maps: Post floor plans or maps near entrances and exits, clearly indicating the location of emergency exits, assembly points, and emergency equipment.
- Fire Extinguisher and Safety Equipment Signage: Clearly mark the locations of fire extinguishers, fire alarm pull stations, eyewash stations, safety showers, and first-aid kits with appropriate signage.
- Evacuation Plan: Develop and communicate an evacuation plan for the laboratory or workshop, outlining procedures and assembly points for all occupants.
- Regular Inspection and Maintenance: Regularly inspect and maintain all emergency exit signs, ensuring that they are in good working condition and clearly visible.
- Employee Training: Provide training to all laboratory and workshop personnel on the location of emergency exits, evacuation procedures, and the proper use of emergency equipment.
- Compliance With Regulations: Ensure that all emergency exits and signage comply with local building and safety codes, as well as any specific regulations applicable to laboratories and workshops.

By implementing clear and effective emergency exits and signage, laboratories and workshops can enhance the safety and preparedness of their occupants in the event of an emergency. Regular drills and practice evacuations can further reinforce the importance of emergency procedures and help occupants respond swiftly and calmly during real emergencies.

8.5.3 Chemical Safety Equipment

Chemical safety equipment is crucial in any laboratory where hazardous chemicals are used, handled, or stored. The following are some of the essential chemical safety equipment that every laboratory should have:

a) Fume hoods and ventilation systems: Fume hoods and ventilation systems are critical components in both laboratories and workshops to control and remove airborne contaminants, vapors, fumes, and hazardous gases. These systems play a crucial role in maintaining a safe and healthy work environment by preventing exposure to harmful substances.

i. Fume Hoods
    - Purpose: Fume hoods are enclosed, ventilated workspaces designed to capture and contain hazardous fumes and vapors generated during various laboratory processes.
    - Function: A fume hood typically consists of a front sash or sliding panel that can be opened or closed to allow access to the work area. When the sash is closed, the fumes are effectively contained and exhausted out of the laboratory through ductwork.
    - Types: There are several types of fume hoods, including standard fume hoods, ductless fume hoods (filtered hoods), and specialized hoods for specific applications (e.g., radioisotope hoods and perchloric acid hoods).
    - Safety Measures: Proper use, regular maintenance, and appropriate certification are essential to ensure the effectiveness and safety of fume hoods. Users should follow specific procedures and guidelines to minimize exposure to hazardous materials.

ii. Ventilation Systems
    - Purpose: Ventilation systems are designed to provide a continuous flow of fresh air and remove airborne contaminants from a workspace.
    - Local Exhaust Ventilation (LEV): LEV captures contaminants at or near the source (e.g., fume hoods and dust collectors) and directs them away from the breathing zone of workers.
    - Dilution Ventilation: Dilution ventilation involves providing a constant flow of fresh air to dilute and reduce

airborne contaminants in a larger area. This is commonly used in workshops where localized containment is not practical.
- General Ventilation: General ventilation systems ensure proper air circulation throughout the entire laboratory or workshop, reducing the concentration of airborne contaminants.

Regular maintenance and inspection of fume hoods and ventilation systems are essential to ensure their continued effectiveness. Laboratory personnel and workshop operators should receive training on the proper use of fume hoods and ventilation systems, as well as emergency procedures in case of ventilation system failure or exposure to hazardous materials.

b) Chemical storage cabinets: Chemical storage cabinets are specialized storage units designed to safely store hazardous chemicals in laboratories and workshops. They are crucial for preventing accidents, spills, and potential chemical reactions that could pose risks to personnel and the environment. Here are some key features and considerations for chemical storage cabinets in labs and workshops:

- Material: Chemical storage cabinets are typically made from durable and chemically resistant materials such as metal, polyethylene, or wood with a chemical-resistant coating.
- Ventilation: Many chemical storage cabinets are equipped with built-in ventilation systems or vents to ensure proper airflow and prevent the buildup of hazardous vapors.
- Segregation: Cabinets should have separate compartments or shelves to store different types of chemicals, preventing incompatible substances from coming into contact and reacting.
- Flammable Cabinets: Flammable liquids should be stored in flammable storage cabinets, which are designed to resist fire and prevent the spread of flames in case of a fire inside the cabinet.
- Corrosive Cabinets: Corrosive chemicals, such as acids and bases, should be stored in cabinets made of acid-resistant materials and equipped with acid-resistant shelves and liners.

- Labeling: Proper labeling on the outside of the cabinets is essential to identify the contents and hazards of the stored chemicals.
- Fire Safety: Chemical storage cabinets should be located away from potential ignition sources and emergency exits to minimize fire risks.
- Security: To prevent unauthorized access, cabinets should be lockable, and access should be restricted to authorized personnel only.
- Spill Containment: Cabinets should have spill containment features, such as a leak-proof base or tray, to contain and prevent spills from spreading outside the cabinet.
- Compliance: Chemical storage cabinets should comply with local regulations, safety codes, and standards to ensure that they meet the required safety standards.
- Capacity: Cabinets should have adequate storage capacity to accommodate the quantities of chemicals used in the laboratory or workshop.
- Inspection and Maintenance: Regular inspection and maintenance of the cabinets are crucial to ensure their integrity and functionality over time.

It is essential to train laboratory and workshop personnel on proper chemical storage practices, including organizing, segregating, and labeling chemicals appropriately. Ensuring that chemicals are stored correctly in dedicated chemical storage cabinets is a vital step in maintaining a safe and compliant work environment.

c) Spill kits: Spill kits are essential in both laboratories and workshops to promptly and safely manage spills of hazardous materials. Spills can pose a risk to health, safety, and the environment, so having appropriate spill kits readily available is crucial for effective response. Spill kits are typically designed to handle specific types of spills and may contain various absorbent materials, tools, and PPE. Here are some common types of spill kits for labs and workshops:

i. Chemical Spill Kits
   - Absorbents: Chemical spill kits contain absorbent materials such as spill pads, spill pillows, or absorbent powders designed to safely absorb and contain chemical spills.

- PPE: These kits may include appropriate PPE, such as gloves, safety goggles, and chemical-resistant aprons, to protect personnel during the cleanup process.

ii. Oil Spill Kits
- Absorbents: Oil spill kits typically contain absorbents like oil-only pads, rolls, or booms that are specially designed to absorb oil and repel water.
- Containment: Oil spill kits may also include materials like oil absorbent socks or booms to help contain and prevent the spread of spilled oil.

iii. Universal Spill Kits
- Universal spill kits are designed to handle a variety of spills, including water-based liquids, oils, and some hazardous chemicals.
- They may contain a combination of universal absorbents, PPE, and cleanup tools suitable for a broad range of spills.

iv. Biohazard Spill Kits
Biohazard spill kits are used in laboratories handling biohazardous materials and may include absorbents, disinfectants, and biohazard waste bags for proper disposal of contaminated materials.

v. Mercury Spill Kits
Mercury spill kits are specifically designed to handle spills of elemental mercury, with tools and materials to safely collect and contain the mercury droplets.

vi. Battery Acid Spill Kits
Battery acid spill kits contain neutralizing agents, absorbents, and PPE suitable for handling spills of battery acid or other corrosive substances.

It is essential to ensure that spill kits are placed in easily accessible locations throughout the laboratory or workshop, and personnel should be trained on their proper use. Additionally, regular inspection and maintenance of spill kits are necessary to ensure that the absorbents and materials are in good condition and have not expired. Prompt and proper cleanup of spills can help prevent accidents, minimize damage, and protect the health and safety of all individuals in the workspace.

d) Eyewash stations and safety showers: Stations and safety showers are critical safety equipment in both laboratories and workshops to provide immediate first aid in the event of eye or body exposure to hazardous materials. They offer a quick means to rinse off contaminants and chemicals to prevent further injury. Here are the key features and considerations for eyewash stations and safety showers in lab and workshop settings:

i. Eyewash Stations
    - Location: Eyewash stations should be easily accessible, within a 10-s reach from areas where hazardous materials are used or stored. They should be placed on the same level as the potential hazard to ensure quick access.
    - Flow Rate: Eyewash stations should provide a controlled flow of water to rinse the eyes gently, typically at a minimum flow rate of 0.4 gallons per minute (1.5 liters per minute).
    - Temperature: The water in the eyewash station should be tepid (between 60°F and 100°F or 16°C to 38°C) to prevent further injury to the eyes.
    - Activation: Eyewash stations should have hands-free activation, such as a foot pedal or pull-down handle, to allow the injured person to use the station without needing to touch anything with their hands.

ii. Safety Showers
    - Location: Safety showers should also be easily accessible, within a 10-s reach from areas where significant chemical spills or full-body exposure to hazardous materials can occur.
    - Flow Rate: Safety showers should provide a high flow rate of water to drench the affected person quickly, typically at least 20 gallons per minute (75.7 liters per minute).
    - Temperature: Similar to eyewash stations, safety showers should deliver tepid water to minimize thermal shock.
    - Activation: Safety showers should have a pull-down handle or push plate for hands-free activation.

iii. Inspections and Maintenance
    - Regular inspections of eyewash stations and safety showers are crucial to ensure that they are functioning correctly and free from any obstructions that may hinder their use in emergencies.

- Weekly activation checks and annual servicing are common practices to ensure their reliability.

iv. Training and Awareness
- All personnel in the lab or workshop should receive training on the location, proper use, and maintenance of eyewash stations and safety showers.
- Awareness of these emergency fixtures should be regularly emphasized to ensure that everyone knows their locations and how to access them quickly.

Eyewash stations and safety showers are indispensable safety features that can minimize the impact of accidental exposure to hazardous materials. Ensuring proper installation, maintenance, and training on their use are essential for creating a safe working environment in laboratories and workshops.

e) Labeling of chemicals: Labeling of chemicals is a critical safety practice in both laboratories and workshops. Proper labeling helps ensure that everyone working in these environments can identify the contents of containers and understand the associated hazards. The labeling requirements are often regulated by national or international standards, such as the Globally Harmonized System (GHS). Here are some essential aspects of labeling chemicals in labs and workshops:

- Name and Identity: Each chemical container should be clearly labeled with the name or identity of the substance it contains. The name used should match the name provided in the Safety Data Sheet (SDS) for the chemical.
- Hazard Pictograms: Chemical labels should include hazard pictograms according to the GHS. These pictograms provide a quick visual indication of the primary hazards associated with the chemical.
- Signal Words: Signal words, such as "Danger" or "Warning," are used on chemical labels to convey the severity of the hazards. "Danger" is used for more severe hazards, whereas "Warning" indicates moderate hazards.
- Hazard Statements: Hazard statements describe the nature and degree of the hazard posed by the chemical. These statements should be clear, concise, and directly related to the identified hazards.

- Precautionary Statements: Precautionary statements provide information on how to handle, store, and dispose of the chemical safely. They also outline specific actions to take in case of exposure or accidental release.
- Supplier Information: Chemical labels should include the name and address of the supplier or manufacturer responsible for the chemical.
- Product Identifier: The label should include a unique product identifier that corresponds to the identifier on the SDS.
- Batch or Lot Number: If applicable, a batch or lot number should be included on the label to track specific production batches.
- Expiration Date: If the chemical has a limited shelf life, then the label should indicate an expiration date or date of manufacture.
- Container Integrity: Ensure that labels remain legible and securely attached to the containers. If the label becomes damaged or illegible, then a new label should be affixed.
- Secondary Containers: When transferring chemicals to secondary containers, the secondary containers should also be properly labeled with at least the name of the chemical and any relevant hazards.

Proper chemical labeling promotes safe handling, storage, and disposal of hazardous materials in laboratories and workshops. It is essential to follow the specific labeling requirements outlined in the applicable regulations and standards to maintain a safe working environment for all personnel. Proper handling, storage, and disposal of hazardous chemicals and regular training and communication among lab personnel are critical to maintaining a safe working environment.

8.5.4 Electrical Safety Equipment

Electrical safety equipment is crucial in any laboratory where electrical equipment is used, as electrical hazards can cause serious injury or even death. The following are some of the essential electrical safety equipment that every laboratory should have:

a) Ground-fault circuit interrupters (GFCIs): These are electrical safety devices designed to protect people from electric shocks and electrical fires caused by ground faults. In both laboratories and workshops, GFCIs are essential for ensuring the safety of electrical systems and preventing potential accidents. Here is how GFCIs are used in lab and workshop settings:

i. Lab Use:
   - Electrical Equipment Safety: In laboratories, where various electrical devices and equipment are used, GFCIs are installed to protect researchers and lab personnel from electric shocks in case of a ground fault in any equipment.
   - Wet and Damp Environments: Laboratories often deal with liquids and moisture, which can increase the risk of electrical hazards. GFCIs are particularly important in areas where wet or damp conditions are common, such as near sinks, fume hoods, or water baths.

ii. Workshop Use:
   - Power Tools and Equipment: Workshops use a variety of power tools and equipment, increasing the risk of electrical accidents. GFCIs are installed on outlets to provide an extra layer of safety and protect users from shocks when using power tools.
   - Outdoor Workshops: For workshops located outdoors or in areas exposed to weather conditions, GFCIs are crucial due to the increased risk of electrical hazards caused by wet or damp environments.

iii. Key Benefits of GFCIs in Both Settings:
   - Shock Protection: GFCIs constantly monitor the flow of electrical current and can quickly shut off power when an imbalance is detected, preventing potential electric shocks.
   - Fire Prevention: GFCIs can help prevent electrical fires that may occur due to ground faults or short circuits.
   - Code Compliance: Many building codes and electrical safety standards require the installation of GFCIs in areas prone to electrical hazards, including labs and workshops.

iv. Where to Install GFCIs:
   - Outlets near sinks or water sources in labs and workshops.
   - Outlets in areas exposed to moisture or dampness.
   - Outlets used for power tools and equipment.

It is essential to have GFCIs regularly tested and maintained to ensure that they are functioning correctly. GFCIs should be inspected, tested, and reset periodically as part of electrical safety procedures in both laboratory and workshop environments. Additionally, educating lab personnel and workshop workers about the importance of GFCIs and electrical safety practices is crucial for promoting a safe working environment.

b) Circuit breakers: Circuit breakers are important safety devices used in both laboratories and workshops to protect electrical circuits and equipment from overcurrent and short circuits. They are designed to interrupt the flow of electricity when an abnormal condition is detected, preventing damage to electrical devices and reducing the risk of electrical fires. Here is how circuit breakers are used in labs and workshops:

- Circuit Protection: In both labs and workshops, circuit breakers are installed in the electrical distribution panels to protect individual circuits. When an excessive current flows through a circuit, such as during a short circuit or overload, the circuit breaker trips and interrupts the current flow, cutting off the electricity to the affected circuit.
- Preventing Electrical Fires: Circuit breakers play a critical role in preventing electrical fires caused by overloads or faults. By quickly interrupting the current flow, they help avoid overheating of wires and components that could lead to fires.
- Resettable Protection: Unlike fuses, circuit breakers are resettable. After they trip due to an overcurrent event, they can be manually reset by toggling the breaker switch back to its "ON" position once the issue is resolved.
- Different Types of Circuit Breakers: There are different types of circuit breakers available, each designed for specific applications and current ratings. Common types include thermal-magnetic circuit breakers, electronic circuit breakers, and residual current circuit breakers used for ground-fault protection.
- Circuit Breaker Ratings: In both labs and workshops, the appropriate circuit breaker rating is determined based on the electrical load of the circuit it protects. It is essential to ensure that the circuit breaker's rating matches the electrical load it will encounter.

- Electrical Safety: Regular inspection and maintenance of circuit breakers are crucial to ensure their proper functioning. Faulty or malfunctioning circuit breakers should be replaced promptly to maintain electrical safety.
- Electrical Panels: In larger labs and workshops, multiple circuit breakers are housed in electrical panels. These panels distribute electricity to various circuits throughout the facility and allow for easy control and monitoring of electrical systems.

It is important for lab and workshop personnel to be familiar with the location and operation of circuit breakers in their respective areas. Understanding electrical safety and the proper handling of circuit breakers can help prevent electrical accidents and ensure the safe operation of electrical systems in laboratories and workshops.

c) **Insulated tools and equipment:** Insulated tools and equipment are specially designed to provide electrical insulation, protecting users from electrical shocks while working on or near live electrical systems. They are commonly used in both laboratories and workshops where electrical work is performed. Here are some examples of insulated tools and equipment used in these settings:

- Insulated Screwdrivers: Insulated screwdrivers have handles made of non-conductive materials, such as rubber or plastic, to provide protection against electric shock when working on live circuits.
- Insulated Pliers: Insulated pliers have insulated handles to prevent the transmission of electric current through the tool to the user.
- Insulated Adjustable Wrenches: These wrenches have insulated handles to provide protection when adjusting or loosening electrical connections.
- Insulated Wire Strippers: Insulated wire strippers have handles made of non-conductive materials to protect against electric shock when stripping wires.
- Insulated Electrical Gloves: Insulated gloves are designed to provide electrical insulation for hands and are essential when working on live electrical systems.

- Insulated Electrical Mats: Insulated mats are placed on the floor around electrical equipment to provide a safe working surface and prevent electrical shocks.
- Insulated Voltage Testers: These are handheld devices used to check for the presence of voltage in electrical circuits without making direct contact with live parts.
- Insulated Ladders: Insulated ladders are designed to provide electrical insulation for users when working at heights near electrical systems.
- Insulated Extension Cords: Insulated extension cords have protective coverings to prevent accidental contact with live wires.

It is essential to use properly rated and certified insulated tools and equipment, as well as to inspect them regularly for any signs of damage or wear. Following safety protocols and using the right insulated tools can significantly reduce the risk of electrical accidents in laboratories and workshops. Additionally, workers should receive training on the correct use and maintenance of insulated tools and equipment to ensure their safety while performing electrical tasks.

d) Electrical safety signs and labels: Electrical safety signs and labels are essential in both laboratories and workshops to warn personnel of potential electrical hazards and promote a safe working environment. These signs and labels provide visual cues and information about electrical equipment, circuits, and potential dangers. Here are some common electrical safety signs and labels found in labs and workshops:

i. Electrical Hazard Signs:
   - "High Voltage": Indicates areas with high voltage electrical equipment or circuits, warning individuals to stay away or take necessary precautions.
   - "Danger - Electrical Hazard": Alerts personnel to the presence of electrical hazards that could cause injury or death.

ii. Electrical Equipment Labels:
   - "Warning - Disconnect Power Before Servicing": Placed on electrical equipment to remind users to turn off power before conducting maintenance or repairs.

- "Caution - Risk of Electric Shock": Used on equipment that poses a risk of electric shock, prompting users to exercise caution.

iii. Voltage Rating Labels:
- "120V," "240V," etc.: Indicates the voltage rating of electrical outlets or equipment, informing users about the voltage present.

iv. Grounding and Earthing Labels:
- "Earth Ground" and "Grounding Point": Identifies locations where electrical equipment or circuits should be grounded for safety.

v. Electrical Panel Labels:
- "Electrical Panel - Authorized Personnel Only": Indicates that only trained or authorized personnel should access electrical panels.

vi. Circuit Breaker Labels:
- "Circuit Breaker" and "Fuse Box": Labels the location of circuit breakers or fuse boxes for quick identification during emergencies.

vii. Outlet and Switch Labels:
- "On/Off" and "Receptacle": Provides clear markings for electrical outlets and switches.

viii. Cable and Wire Labels:
- "Live Wire," "Neutral," and "Ground": Labels wires and cables to indicate their functions and potential hazards.

ix. Emergency Shut-Off Labels:
- "Emergency Shut-Off": Identifies the location of emergency shut-off switches for immediate power disconnection in case of emergencies.

x. Restricted Access Labels:
- "No Unauthorized Entry" and "Keep Out": Warns unauthorized personnel from accessing areas with electrical hazards or sensitive equipment.

These signs and labels should be easily visible and located in appropriate positions, especially near electrical equipment and panels. They should have clear and concise messaging using recognizable symbols and standardized colors (e.g., black text on a yellow background for warning signs). Regular inspection and maintenance of electrical safety signs and labels are crucial to ensure their legibility and effectiveness in preventing accidents and injuries related to electrical hazards in the lab and workshop.

8.5.5 Biological Safety Equipment

Biological safety equipment is crucial in any laboratory where biological materials are handled or processed, as these materials can pose a hazard to the laboratory personnel and the environment. The following are some of the essential biological safety equipment that every laboratory should have:

a) Biosafety cabinets: Biosafety cabinets are specialized containment devices designed to provide a safe working environment when handling hazardous biological materials, such as microorganisms or biohazardous substances. They are essential in both laboratories and workshops where biological research or manipulations are conducted. Biosafety cabinets help protect both laboratory personnel and the surrounding environment from potential exposure to biological agents. There are different types of biosafety cabinets, classified based on their design and level of containment:

i. Class I Biosafety Cabinets:
   - Airflow: Class I cabinets provide personnel protection through a unidirectional airflow that is drawn from the front of the cabinet, passing over the work area, and exhausted through a high efficiency particulate air (HEPA) filter.
   - Containment: They offer minimal protection to the environment but protect personnel from inhaling harmful aerosols or particles.
   - Usage: Class I biosafety cabinets are typically used for work involving low to moderate-risk agents, such as diagnostic procedures or sample manipulations.

ii. Class II Biosafety Cabinets:
   - Types: Class II cabinets are further divided into four types: A1, A2, B1, and B2, each with different airflow patterns and levels of protection.
   - Airflow: These cabinets provide both personnel and environmental protection through HEPA-filtered supply and exhaust air.
   - Containment: Class II cabinets are suitable for working with moderate to high-risk biological agents, providing a barrier against potential contamination.
   - Usage: They are commonly used in research laboratories, diagnostic facilities, and biotechnology applications.

iii. Class III Biosafety Cabinets (Glove Boxes):
   - Airflow: Class III cabinets are totally enclosed, gas-tight systems with a glove box configuration. All operations are conducted through attached gloves, and the cabinet maintains negative pressure to ensure containment.
   - Containment: Class III cabinets offer the highest level of containment and are designed for handling highly hazardous pathogens and agents.
   - Usage: Class III biosafety cabinets are used in high-containment laboratories and are suitable for working with dangerous pathogens and viruses.

It is essential to choose the appropriate class and type of biosafety cabinet based on the risk assessment of the biological materials being handled in the laboratory or workshop. Regular maintenance, certification, and adherence to standard operating procedures are necessary to ensure the effective performance of biosafety cabinets and to maintain a safe working environment when dealing with biohazards.

b) Autoclaves: Autoclaves are essential equipment used in both laboratories and workshops for sterilization purposes. They are designed to apply high-pressure steam to destroy microorganisms, including bacteria, viruses, and spores, thus ensuring that lab equipment and certain materials are free from contamination. The use of autoclaves helps maintain a sterile environment and prevent the spread of infectious agents. Here is how autoclaves are used in both settings:

i. In Laboratories:
   - Sterilization of Lab Equipment: Autoclaves are used to sterilize various laboratory equipment, such as glassware, pipettes, petri dishes, and culture media.
   - Decontamination of Biological Waste: Autoclaves are used to decontaminate biological waste, such as used agar plates, pipette tips, and culture materials.
   - Media Preparation: Autoclaves are used to prepare culture media by sterilizing them before use, ensuring a clean environment for culturing microorganisms.
   - Biohazard Waste Treatment: Autoclaves are used to treat biohazardous waste generated in laboratories,

including items contaminated with potentially harmful biological agents.

ii. In Workshops:
   - Sterilization of Tools: Autoclaves are used in workshops, particularly in medical and dental settings, to sterilize surgical instruments and dental tools.
   - Cleaning and Sterilization of Lab Coats and PPE: In certain workshops, autoclaves may be used to clean and sterilize lab coats, gowns, and PPE for reuse.
   - Sterilization of Packaging Materials: Autoclaves may be used to sterilize packaging materials for medical devices or sensitive electronic components.
   - Testing and Quality Control: Autoclaves may be used in workshops to test the performance and sterilization efficiency of certain products.

In both laboratories and workshops, autoclaves are vital tools for maintaining a safe and clean environment. Regular maintenance, calibration, and validation of autoclaves are necessary to ensure their proper functioning and accuracy in achieving effective sterilization. Proper training on autoclave operation and safety protocols is crucial for users to handle the equipment correctly and to prevent accidents or exposure to high-pressure steam.

c) Proper waste disposal systems: Proper waste disposal systems are essential in both laboratories and workshops to ensure the safe and responsible management of various types of waste generated during daily activities. Implementing appropriate waste disposal systems helps prevent environmental contamination, reduces health risks, and ensures compliance with relevant regulations. Here are some key components of proper waste disposal systems for labs and workshops:

- Waste Segregation: Separate waste at the source into different categories, such as general waste, hazardous waste, recyclables, and electronic waste. Lab and workshop personnel should be trained on the proper segregation procedures.
- Clearly Labeled Bins: Provide clearly labeled bins or containers for different waste categories to facilitate proper segregation. Use color-coded bins where applicable to improve recognition.

- Hazardous Waste Containers: For laboratories, use specifically designed containers for hazardous waste, such as chemicals, biohazardous materials, and broken glassware. Ensure these containers are leak-proof and appropriately labeled.
- Chemical Waste Disposal: Develop protocols for disposing of chemicals safely. Contact a licensed waste disposal service to properly handle and dispose of hazardous chemicals according to local regulations.
- Sharps Containers: Use puncture-resistant sharps containers for the disposal of used needles, syringes, and other sharp objects in both labs and workshops.
- Recycling Stations: Set up recycling stations for the collection of recyclable materials, such as paper, cardboard, plastics, and glass, in both labs and workshops.
- E-Waste Collection: Implement a system for the proper disposal of electronic waste, such as old computers, printers, and other electronic equipment, to prevent environmental contamination.
- Waste Collection Schedule: Establish a regular waste collection schedule to prevent accumulation and maintain a clean and safe environment.
- Training and Awareness: Provide training to lab and workshop personnel on waste disposal procedures, emphasizing the importance of proper waste management for safety and environmental protection.
- Compliance with Regulations: Stay informed about local, state, and federal regulations governing waste disposal and ensure that your waste management practices align with these requirements.
- Waste Audits: Conduct periodic waste audits to assess the effectiveness of the waste disposal system, identify areas for improvement, and ensure compliance with waste disposal policies.

It is important to note that the above equipment is only a part of a comprehensive biological safety program. By implementing a well-designed waste disposal system and fostering a culture of responsible waste management, labs and workshops can contribute to a safer and more environmentally sustainable working environment.

8.5.6 Radiation Safety Equipment

Radiation safety equipment is crucial in any laboratory where radiation sources are used or handled, as exposure to radiation can cause serious injury or even death. The following are some of the essential radiation safety equipment that every laboratory should have:

a) Lead shielding: Lead shielding is used in laboratories and workshops to protect individuals from exposure to radiation or to contain sources of radiation safely. The heavy density of lead makes it an effective material for attenuating ionizing radiation, such as gamma rays and X-rays. Here are some common applications of lead shielding in labs and workshops:

- Radiology and X-ray Rooms: In medical facilities and research labs that use X-ray machines, lead shielding is essential to protect patients, medical staff, and other individuals from unnecessary radiation exposure. Lead aprons and lead-lined walls or barriers are commonly used for this purpose.
- Nuclear Medicine Labs: In nuclear medicine departments or research facilities, lead shielding is used to store and handle radioactive materials safely. Lead containers and cabinets are used to minimize the risk of radiation exposure during handling and storage.
- Particle Accelerators and Research Facilities: High-energy particle accelerators and research facilities dealing with radioactive isotopes often use lead shielding to protect personnel from potential radiation hazards.
- Industrial X-ray Testing: In industrial settings where X-ray testing is performed to inspect materials or products, lead shielding is used to protect workers and ensure radiation safety.
- Radiotherapy Treatment Rooms: In cancer treatment centers, lead shielding is used in radiation therapy rooms to protect healthy tissues and organs surrounding the tumor being treated.
- Nuclear and Radiochemistry Labs: Laboratories that handle radioactive materials or conduct radiochemical research may use lead shielding to reduce the risk of radiation exposure to personnel.

- Radioactive Waste Storage: Lead containers or casks are used to safely store and transport radioactive waste materials, minimizing the potential for radiation release.

Lead shielding is available in various forms, such as lead bricks, lead sheets, lead glass, and lead-lined walls or rooms. The thickness and configuration of the lead shielding are tailored to the specific radiation source and level of protection required. It is essential to follow appropriate safety guidelines and regulations when using lead shielding to ensure adequate protection and prevent unnecessary radiation exposure in laboratories and workshops. Regular monitoring and inspection of the lead shielding are also necessary to ensure its continued effectiveness.

b) Geiger counters: Geiger counters, also known as Geiger-Muller counters, are radiation detection devices used to measure ionizing radiation levels. They detect the presence of alpha, beta, and gamma radiation and are commonly used in both laboratories and workshops where radioactive materials may be handled or stored.

i. Laboratory Applications:
   - Radiation Research: Geiger counters are used in laboratories engaged in radiation research and experiments to measure radiation levels and analyze the properties of radioactive materials.
   - Nuclear Medicine: In medical laboratories, Geiger counters are used to monitor and measure radioisotopes used in diagnostic imaging and treatment, such as gamma-emitting isotopes.
   - Radioactive Contamination Monitoring: Geiger counters are used to detect and monitor potential radioactive contamination in lab areas, equipment, or personnel.

ii. Workshop Applications:
   - Radioactive Source Detection: In some workshops, radioactive materials may be used for specific purposes, such as industrial radiography. Geiger counters are used to check for the presence of radioactive sources and ensure safety.
   - Radiography and Non-Destructive Testing: Geiger counters can be used to assess the levels of radiation exposure during radiography and other non-destructive testing procedures.

iii. Key features of Geiger counters include the following:
   - Detection Sensitivity: Geiger counters can detect low levels of radiation, making them valuable for monitoring even small amounts of radioactive materials.
   - Audible and Visual Indicators: Geiger counters typically produce audible clicks or beeps and visual displays to indicate the presence and intensity of radiation.
   - Portable: Many Geiger counters are handheld and portable, allowing for easy use in different areas of the lab or workshop.
   - Battery-Powered: Geiger counters are often powered by batteries, ensuring that they can be used even in areas without a power source.
   - Measurement Units: Geiger counters measure radiation in units such as counts per minute or counts per second and can sometimes display radiation dose rates in millirem or millisieverts.

iv. Safety Considerations:
   - Proper Training: Users of Geiger counters should receive proper training in their operation, maintenance, and safety protocols for handling radioactive materials.
   - Calibration and Maintenance: Regular calibration and maintenance of Geiger counters are essential to ensure accurate readings.
   - Shielding: Adequate shielding should be used when handling or measuring radioactive materials to reduce radiation exposure.
   - Regulatory Compliance: Use of Geiger counters should comply with applicable regulations and guidelines for handling and working with radioactive materials.

By employing Geiger counters in laboratories and workshops, workers can monitor radiation levels, detect potential hazards, and implement appropriate safety measures to protect themselves and the environment.

c) Dosimeters: Dosimeters are devices used to measure and record the cumulative amount of radiation exposure received by individuals over a specific period. In both laboratories and workshops where radiation hazards may be present, dosimeters play a critical role in monitoring and ensuring the safety of personnel. There are different types of dosimeters used in these environments:

i. Personal Dosimeters:
    - These dosimeters are worn by individuals who may be exposed to radiation during their work. They are usually small and lightweight, designed to be worn on clothing or placed in a pocket close to the body.
    - Personal dosimeters continuously measure and record the radiation dose received by the wearer, providing real-time data on radiation exposure levels.
    - Different types of personal dosimeters include film badges, thermoluminescent dosimeters (TLDs), optically stimulated luminescence dosimeters (OSLDs), and electronic personal dosimeters (EPDs).

ii. Area Dosimeters:
    - Area dosimeters are placed in specific locations in the laboratory or workshop to monitor radiation levels in different areas.
    - These dosimeters provide valuable data on the distribution of radiation in the working environment and help assess potential exposure risks.

iii. Electronic Dosimeters:
    - Electronic dosimeters are sophisticated devices that offer real-time readings of radiation exposure. They can display dose rates, accumulated dose, and other relevant information.
    - Some electronic dosimeters are equipped with alarms or warning indicators to alert the wearer when radiation levels exceed safe limits.

iv. Passive Dosimeters:
    - Passive dosimeters do not require a power source or batteries to operate. They rely on the detection of radiation to record exposure data.
    - Film badges, TLDs, and OSLDs are examples of passive dosimeters commonly used in laboratories and workshops.

The selection of dosimeters depends on the specific radiation sources present, the nature of work, and the regulations and guidelines applicable to the facility. Regular calibration, maintenance, and monitoring of dosimeters are crucial to ensuring their accuracy and reliability. Additionally, dosimeter data is used to assess compliance with safety standards and to make informed decisions about radiation protection measures in the laboratory or workshop environment.

d) Radiation warning signs and labels: Radiation warning signs and labels are essential in both laboratories and workshops where radiation sources are present. These signs and labels serve to alert personnel and visitors about potential radiation hazards and provide important safety information. The specific design and content of radiation warning signs and labels may vary depending on the type of radiation and the regulatory requirements. Here are some common types of radiation warning signs and labels:

- Caution (Cautionary) Signs: Caution signs are typically yellow with a black radiation symbol (trefoil) and text. They warn of the presence of a radiation source and provide basic information about the type of radiation and any precautions to take.
- Warning Signs: Warning signs are typically red with a black radiation symbol and text. They indicate a more significant radiation hazard that requires heightened caution and adherence to safety protocols.
- Radioactive Material Labels: Radioactive material labels are affixed to containers holding radioactive substances. They include the radioactive symbol, radioisotope name, and activity, as well as information about handling precautions.
- Laser Radiation Signs: In areas where lasers are used, laser radiation signs indicate the presence of laser radiation and provide information about laser class, potential hazards, and required protective measures.
- X-ray Radiation Signs: X-ray radiation signs are used in areas where X-ray equipment is present. They warn of X-ray radiation hazards and may include specific safety instructions for using the equipment.
- Emergency Shutdown Signs: Emergency shutdown signs are placed near emergency switches or controls to indicate their function in quickly shutting down radiation sources in case of an emergency.

- Restricted Access Signs: Restricted access signs are used to designate areas where only authorized personnel are allowed due to potential radiation hazards.
- Dosimetry Area Signs: Dosimetry area signs indicate locations where personnel wear dosimeters to monitor radiation exposure levels.
- Radiation Area Signs: Radiation area signs are used to mark areas where radiation levels exceed certain limits and require special precautions.

It is important to note that radiation safety equipment alone is insufficient to ensure laboratory safety. Proper handling, storage, and disposal of radioactive materials and regular training and communication among lab personnel are also critical to maintaining a safe working environment. Additionally, it is important to follow appropriate radiation safety guidelines, such as those established by the Nuclear Regulatory Commission, to prevent exposure to ionizing radiation.

8.5.7 Emergency Response Equipment

Emergency response equipment is crucial in any laboratory, as accidents or incidents can occur at any time. The following are some of the essential emergency response equipment that every laboratory should have:

a) First-aid kits: First-aid kits are essential in both laboratories and workshops to provide immediate medical assistance for minor injuries or incidents. The contents of first-aid kits may vary based on the specific needs and hazards of each environment, but they typically include a variety of supplies to treat common injuries. Here are some common items found in first-aid kits for labs and workshops:

- Adhesive Bandages (Assorted Sizes): For treating small cuts, abrasions, and minor wounds.
- Sterile Gauze Pads and Dressings: Used for larger wounds to stop bleeding and protect the injury.
- Adhesive Tape: To secure dressings and bandages in place.
- Antiseptic Wipes or Solution: For cleaning wounds and preventing infection.
- Disposable Gloves: To protect the first aid provider and the injured person from potential contamination.

- Scissors and Tweezers: For cutting bandages and removing splinters or foreign objects.
- Instant Cold Packs: To reduce swelling and relieve pain in case of sprains or strains.
- Triangular Bandage: Can be used as a sling or for creating pressure dressings.
- Cardiopulmonary Resuscitation (CPR) Face Shield or Pocket Mask: For performing CPR safely.
- Eye Wash Solution: For rinsing eyes in case of chemical splashes or foreign objects.
- Burn Gel or Dressings: To treat minor burns and scalds.
- Resealable Plastic Bags: For disposing of contaminated materials or storing used items.
- First-Aid Manual: Instructions on how to provide basic first aid.
- Emergency Contact Information: Contact numbers for emergency services and medical facilities.

It is essential to regularly check and replenish the first-aid kit to ensure that all supplies are up to date and available for use in emergencies. Additionally, first-aid kits should be easily accessible and located in areas where they can be quickly retrieved in case of an injury. In some workplaces or labs with specific hazards, additional items such as eyewash stations, specialized burn dressings, or other medical supplies may be necessary to address unique risks. The contents of the first-aid kit should be tailored to the specific needs of the environment and the potential injuries that may occur.

b) Emergency eyewash stations and safety showers: Emergency eyewash stations and safety showers are essential safety equipment in both laboratories and workshops to provide immediate relief and treatment in the event of chemical or hazardous material exposure to the eyes or body. These facilities are designed to minimize the impact of chemical splashes, spills, or accidental contact, ensuring the safety and well-being of personnel. Here is a brief overview of each:

i. Emergency Eyewash Stations:
   - Purpose: Emergency eyewash stations are designed to quickly flush the eyes with a continuous flow of water when they come into contact with harmful substances.

They are primarily used to treat chemical splashes or foreign objects in the eyes.
- Activation: The eyewash station is activated by a quick and easy-to-reach handle or foot pedal, allowing the user to initiate the flow of water immediately.
- Location: Emergency eyewash stations are strategically placed in areas where there is a higher risk of eye exposure to hazardous materials, typically within 10 s of the work area.

ii. Safety Showers:
- Purpose: Safety showers provide a large volume of water to rinse off the body in case of chemical spills or exposure to hazardous substances. They are used to treat skin and body contamination and are especially crucial for situations where clothing may be contaminated.
- Activation: Safety showers are activated by a pull handle or a lever, enabling the user to access the water flow promptly.
- Location: Safety showers are placed in areas where there is a higher risk of body exposure to hazardous materials, typically within 10 s of the work area.

Both emergency eyewash stations and safety showers should comply with relevant safety standards and regulations. Regular inspections, testing, and maintenance are critical to ensure that these safety equipment items are in proper working condition when needed. Training on the proper use and location of emergency eyewash stations and safety showers should be provided to all personnel working in the laboratory or workshop to ensure that they can respond quickly and effectively in case of an emergency.

c) Emergency escape equipment: Emergency escape equipment in laboratories and workshops is essential to ensure the safe evacuation of personnel in the event of an emergency or hazardous situation. The following are some of the commonly used emergency escape equipment in labs and workshops:

- Emergency Exits and Exit Signs: Clearly marked emergency exits with illuminated exit signs provide a clear path for evacuation during emergencies. These exits should be unobstructed and easily accessible.

- Emergency Lighting: Emergency lighting is crucial during power outages or situations where visibility is reduced. It helps guide personnel toward exits and escape routes.
- Fire Alarm System: A fire alarm system, including smoke detectors and heat detectors, alerts occupants to the presence of a fire, triggering the evacuation process.
- Fire Extinguishers: Fire extinguishers should be strategically placed throughout the laboratory or workshop to allow quick access in the event of a small fire that can be controlled.
- Emergency Eyewash Stations: Emergency eyewash stations are crucial in laboratories where there is a risk of chemical splashes to the eyes. They provide immediate irrigation to the eyes in case of exposure to hazardous substances.
- Safety Showers: Safety showers provide a continuous flow of water to rinse off chemicals that come into contact with the skin. They are essential for immediate decontamination in the event of chemical spills or splashes.
- First-Aid Kits: Well-equipped first-aid kits should be readily available to provide immediate medical attention to injured personnel before professional help arrives.
- Personal Gas Monitors: In workshops where hazardous gases or vapors may be present, personal gas monitors can alert workers to dangerous levels of gases, enabling them to evacuate or take appropriate safety measures.
- PPE Storage: Proper storage of PPE, such as safety goggles, face shields, gloves, and lab coats, ensures that workers can quickly don the necessary protective gear during an emergency.
- Communication Systems: Intercoms or two-way radios can facilitate communication during emergencies, allowing for coordination and dissemination of vital information.
- Evacuation Plans and Procedures: Clearly posted evacuation plans and procedures that outline escape routes and assembly points help ensure a safe and orderly evacuation during emergencies.
- Emergency Response Equipment: Specialized emergency response equipment, such as spill kits, decontamination supplies, and radiation monitors (in certain laboratories), should be readily available for specific hazards.

Regular inspection, maintenance, and training on the use of emergency escape equipment are crucial to ensure that personnel can respond effectively and safely in emergencies, reducing the risk of injuries and protecting lives.

d) Emergency lighting and signage: Emergency lighting and signage are essential safety features in laboratories and workshops to ensure that occupants can safely evacuate the premises in the event of a power failure or emergency situation. They help guide people to exits and emergency equipment during low visibility conditions. Here is a breakdown of their importance in labs and workshops:

i. Emergency Lighting:
   - Illumination During Power Outages: In the event of a power failure or other emergencies that cause a loss of regular lighting, emergency lights automatically activate to provide illumination, preventing panic and allowing safe movement.
   - Exit Route Visibility: Emergency lighting ensures that exit routes, stairways, and doorways remain visible during emergencies, assisting occupants in finding their way to safety.
   - Compliance With Safety Regulations: Many safety codes and regulations require the installation of emergency lighting in public and commercial spaces, including laboratories and workshops.
   - Backup Power Source: Emergency lights are often connected to backup power sources, such as batteries or generators, ensuring that they remain operational during power disruptions.

ii. Emergency Signage:
   - Exit Signs: These signs are strategically placed to indicate the location of emergency exits, allowing occupants to quickly find the nearest safe exit during an emergency.
   - Fire Exit Signs: Fire exit signs are specifically designed to guide people to exits that lead to a safe location outside the building in the event of a fire.
   - Safety Equipment Signage: Signage indicating the location of fire extinguishers, fire blankets, first-aid kits,

eyewash stations, and emergency showers helps occupants locate essential safety equipment quickly.

- Regulatory Compliance: Properly placed and visible emergency signage helps organizations comply with safety regulations and building codes.
- Universal Symbols: Emergency signage often uses universally recognized symbols to overcome language barriers and ensure that occupants can understand the signs regardless of their language proficiency.
- Reflective Materials: Some emergency signage is made with reflective materials to enhance visibility in low-light conditions or when illuminated by emergency lighting.

The combination of emergency lighting and signage ensures that in the event of an emergency, such as a fire, hazardous material spill, or power outage, all occupants can quickly and safely evacuate the premises and access necessary safety equipment. Regular maintenance and testing of emergency lighting systems are crucial to ensure that they function correctly when needed. Additionally, emergency evacuation plans should be established, communicated to all occupants, and practiced periodically to familiarize everyone with the proper evacuation procedures.

It is important to note that emergency response equipment alone is not sufficient to ensure laboratory safety. A comprehensive emergency response plan should also be in place that outlines the specific procedures to follow in case of an emergency. The plan should include procedures for evacuating the laboratory, contacting emergency services, and providing first aid or other necessary medical treatment. Regular training and communication among lab personnel are also critical to ensuring that everyone is aware of the emergency response procedures and can respond appropriately in case of an emergency.

8.5.8 General Laboratory Safety Equipment

General laboratory safety equipment is essential in any laboratory to ensure that the work environment is safe for all personnel. The following are some of the essential general laboratory safety equipment that every laboratory should have:

a) Slip-resistant flooring: Slip-resistant flooring is crucial in both laboratories and workshops to reduce the risk of slips, trips, and falls, especially in areas where spills, liquids, or debris are common. Slip-resistant flooring provides better traction underfoot and helps prevent accidents, creating a safer working environment. Here are some common types of slip-resistant flooring used in labs and workshops:

- Epoxy Coatings: Epoxy coatings are durable and provide a smooth, seamless surface with excellent slip resistance. They are often used in laboratories and workshops to create a clean and easily cleanable surface.
- Non-Slip Paint: Non-slip paint is applied on top of existing flooring materials to improve traction. It contains additives like sand or aluminum oxide to create a textured surface.
- Rubber Flooring: Rubber flooring is resilient and offers good slip resistance even when wet. It is suitable for areas with high foot traffic and potential exposure to liquids or chemicals.
- Vinyl Flooring: Some types of vinyl flooring come with slip-resistant properties, making them suitable for areas where spills or moisture may be present.
- Anti-Fatigue Mats: Anti-fatigue mats often have slip-resistant properties and provide cushioning to reduce strain on the legs and back. They are beneficial in areas where employees stand for extended periods.
- Interlocking Tiles: Interlocking floor tiles are easy to install and can be made from rubber or PVC, providing a slip-resistant surface.
- Concrete With Texture: For workshops, textured concrete floors can offer slip resistance, especially when sealed with a non-slip coating.

When choosing slip-resistant flooring for laboratories and workshops, it is essential to consider factors such as the nature of the work, the potential for spills or liquids, and the cleaning requirements. Regular maintenance and cleaning of slip-resistant flooring are necessary to ensure its continued effectiveness in preventing slips and falls. Additionally, ensuring good lighting and clear signage for hazardous areas can further enhance safety in the workspace.

b) Proper storage systems: Proper storage systems are essential in both laboratories and workshops to ensure a safe and organized working environment. Proper storage helps prevent accidents, maintain the integrity of materials, and improve efficiency. Here are some important considerations for storage systems in labs and workshops:

i. Laboratory Storage Systems:
   - Chemical Storage Cabinets: Chemicals should be stored in dedicated chemical storage cabinets that are appropriately labeled and organized based on their compatibility. Flammable, corrosive, toxic, and reactive chemicals should be segregated and stored separately.
   - Flammable Storage Cabinets: Flammable liquids should be stored in flammable storage cabinets designed to prevent the spread of fire and protect against ignition sources.
   - Refrigerators and Freezers: Laboratory-grade refrigerators and freezers should be used for storing temperature-sensitive materials, samples, and reagents. They may have special compartments for hazardous materials.
   - Glassware Storage: Glassware, such as beakers, flasks, and test tubes, should be stored on sturdy racks or shelving units to prevent breakage and keep the lab organized.
   - Biological Safety Cabinets (BSCs): In biological laboratories, BSCs are used to provide containment and protect against biohazards.
   - Microscope Slide Storage: Microscope slides should be stored in slide boxes or racks to prevent damage and contamination.
   - Gas Cylinder Storage: Gas cylinders should be securely stored in a dedicated area, preferably in well-ventilated storage cages or racks, to prevent tipping and damage.

ii. Workshop Storage Systems:
   - Tool Cabinets: Tools should be stored in labeled tool cabinets or tool chests to keep them organized and prevent injuries caused by loose tools.
   - Parts Bins and Organizers: Small parts and components should be stored in labeled bins or organizers to make them easily accessible and prevent loss.
   - Material Storage Racks: Heavy materials or supplies, such as lumber, pipes, and metal sheets, should be

stored on sturdy racks to prevent clutter and potential hazards.

- Pegboards and Tool Walls: Pegboards and tool walls are useful for hanging frequently used tools, ensuring easy access and quick identification.
- Hazardous Material Storage: Any hazardous materials or chemicals in the workshop should be stored in appropriate containers and cabinets, complying with safety regulations.
- Lockable Cabinets: Sensitive equipment, valuable tools, and PPE can be stored in lockable cabinets to ensure security.
- Safety Equipment Storage: Safety equipment, such as fire extinguishers, first-aid kits, and safety goggles, should be easily accessible and stored in designated areas.

Proper storage systems help maintain order, safety, and efficiency in both laboratories and workshops. Regular inspections, labeling, and appropriate training for personnel are essential for effective storage management. Additionally, adherence to safety regulations and guidelines ensures a safe and productive working environment.

c) Well-maintained and regularly inspected equipment: Well-maintained and regularly inspected equipment is crucial for ensuring the safety, efficiency, and reliability of both laboratories and workshops. Proper maintenance and regular inspections help identify potential issues early on, prevent accidents, and extend the lifespan of equipment. Here are the key aspects of maintaining and inspecting equipment in labs and workshops:

- Maintenance Schedule: Establish a maintenance schedule for each piece of equipment, outlining the frequency and type of maintenance required. This schedule should be based on the manufacturer's recommendations, usage intensity, and specific equipment requirements.
- Preventive Maintenance: Conduct regular preventive maintenance to proactively address potential problems. This includes cleaning, lubricating, calibrating, and adjusting equipment as needed.
- Qualified Technicians: Ensure that maintenance and repairs are performed by qualified technicians who are trained

to work with specific equipment types. Regular training and certification programs for maintenance personnel are essential.

- Record Keeping: Maintain detailed records of all maintenance activities, including dates of inspections, repairs, and replacements. These records help track equipment performance, identify trends, and plan for future maintenance needs.
- Safety Checks: Include safety checks as part of the maintenance routine to verify that safety features and mechanisms are functioning correctly.
- Calibration: Regularly calibrate sensitive instruments and equipment to ensure accurate and reliable measurements.
- Lubrication: Keep equipment properly lubricated to reduce friction, prevent wear, and maintain smooth operation.
- Cleaning: Regularly clean equipment to remove dust, debris, and contaminants that can affect performance and accuracy.
- Replacement of Wear Parts: Replace wear parts, such as belts, filters, and seals, at the recommended intervals to prevent equipment failures.
- Inspection Checklist: Develop a detailed inspection checklist specific to each type of equipment, covering critical components and safety features.
- Emergency Shut-Off: Ensure that all equipment has easily accessible emergency shut-off switches or buttons in case of emergencies.
- Training for Users: Provide training to laboratory and workshop personnel on the proper use, care, and maintenance of equipment to minimize the risk of accidents caused by misuse or mishandling.
- Compliance With Regulations: Ensure that all equipment meets relevant safety regulations and standards.
- Regular Audits: Conduct periodic audits to assess the overall safety and maintenance practices within the laboratory or workshop.

By implementing these measures, laboratories, and workshops can maintain a safe and productive working environment while prolonging the life and efficiency of their equipment.

d) Safety signs and labels: Safety signs and labels play a vital role in both laboratories and workshops by providing clear and concise information about potential hazards, safety procedures, and important safety instructions. They are essential for promoting a safe working environment and ensuring that workers and visitors are aware of the risks and precautions to take. Here are some common safety signs and labels used in labs and workshops:

- Hazard Signs: These signs use symbols or text to indicate specific hazards, such as chemical hazards, biological hazards, electrical hazards, and radiation hazards. They serve as warning signs to alert individuals to the presence of potential dangers.
- Fire Extinguisher Signs: These signs indicate the location of fire extinguishers in the lab or workshop, ensuring that people can easily locate them in case of a fire emergency.
- Emergency Exit Signs: These signs indicate the location of emergency exits and evacuation routes, guiding people to safety during an emergency.
- First Aid Signs: First aid signs indicate the location of first-aid kits or stations, providing easy access to medical supplies in case of injuries.
- PPE Signs: These signs remind individuals to wear the appropriate PPE, such as safety goggles, gloves, or hearing protection, when entering specific areas or performing certain tasks.
- Biohazard Signs: Biohazard signs are used in laboratories to identify areas or materials that may contain biohazards, such as infectious agents or potentially hazardous biological materials.
- Chemical Hazard Labels: Chemical hazard labels are affixed to containers of hazardous substances to provide information about the contents, potential hazards, and safety precautions for handling and storage.
- Electrical Hazard Signs: These signs warn individuals of electrical hazards, such as high voltage areas or equipment that poses a risk of electric shock.
- No Smoking Signs: No smoking signs are displayed to remind individuals that smoking is prohibited in certain areas due to fire hazards or the presence of flammable materials.

- Restricted Area Signs: These signs indicate areas that are restricted to authorized personnel only due to safety or security concerns.
- Equipment Safety Labels: Equipment safety labels provide information about safe operation, maintenance, and potential risks associated with specific machinery or tools.
- Slip/Trip/Fall Hazard Signs: These signs highlight areas where there is a risk of slipping, tripping, or falling, alerting individuals to exercise caution.

These signs and labels should be prominently displayed, easy to read, and compliant with relevant safety regulations. Regular inspections should be conducted to ensure that signs and labels remain visible, accurate, and in good condition. Additionally, clear communication and proper training are essential to ensure that individuals understand and follow the safety information provided by these signs and labels.

e) Laboratory safety manuals and training: Laboratory safety manuals and training are essential components of promoting a safe working environment in both laboratories and workshops. They help educate personnel about potential hazards, safe practices, emergency procedures, and the proper use of PPE. Here is an overview of the importance and key elements of laboratory safety manuals and training:

i. Laboratory Safety Manuals:
   - Purpose: Laboratory safety manuals serve as comprehensive guides that outline safety protocols, procedures, and guidelines specific to the laboratory's activities and hazards.
   - Content: The manual should cover topics such as chemical safety, biological safety, fire safety, electrical safety, PPE usage, emergency response procedures, waste disposal guidelines, and equipment safety.
   - Roles and Responsibilities: Clearly define the roles and responsibilities of laboratory personnel, supervisors, and safety officers in maintaining a safe working environment.
   - Compliance: The safety manual should align with relevant safety regulations, local codes, and institutional guidelines.

ii. Laboratory Safety Training:
    - Importance: Safety training is crucial to ensure that all laboratory personnel, including researchers, students, and support staff, are aware of the potential hazards and know how to work safely.
    - Topics: Training sessions should cover specific safety protocols outlined in the safety manual, including chemical handling, emergency response, PPE usage, equipment operation, and waste management.
    - Hands-On Training: Whenever possible, hands-on training should be conducted, allowing participants to practice using safety equipment and responding to simulated emergencies.
    - New Employee Orientation: New employees should undergo comprehensive safety training as part of their onboarding process.
    - Regular Refresher Training: Periodic refresher training sessions should be provided to reinforce safety protocols and address any updates or changes to the safety manual.

iii. Workshop Safety Manuals and Training:
    - Similarities: Workshop safety manuals and training share many similarities with laboratory safety manuals and training, as both aim to protect personnel from workplace hazards.
    - Unique Hazards: Workshops may have specific hazards related to machinery, tools, and heavier materials, requiring specialized safety protocols.
    - Hands-On Training: In workshops, hands-on training is crucial for teaching safe operation and handling of tools and machinery.
    - Specific PPE: Workshop safety training should emphasize the use of appropriate PPE, such as safety glasses, gloves, and hearing protection.

iv. Implementation and Enforcement:
    - Commitment: Effective safety training requires commitment from both management and employees to create a culture of safety.
    - Supervision: Supervisors should ensure that safety protocols are followed and provide ongoing support to personnel.

- Enforcement: Adherence to safety guidelines should be reinforced, and consequences for safety violations should be clearly communicated.

In both laboratory and workshop settings, a proactive approach to safety through comprehensive manuals and training helps prevent accidents, reduces injuries, and creates a safer working environment for all personnel.

8.6 Regular Maintenance and Inspections of Lab Safety Equipment

Regular maintenance and inspections are essential to ensure that laboratory equipment and safety systems function properly and that the work environment remains safe for all personnel. The following are some of the key considerations for regular maintenance and inspections:

a) Importance of regular maintenance: Regular maintenance is of paramount importance in both laboratories and workshops to ensure the safety, efficiency, and longevity of equipment, machinery, and facilities. The significance of regular maintenance can be summarized as follows:

- Safety: Regular maintenance helps identify and address potential safety hazards, minimizing the risk of accidents, injuries, and equipment failures. Faulty or malfunctioning equipment can pose serious dangers to workers in both lab and workshop settings.
- Equipment Performance: Regular maintenance ensures that equipment and machinery function optimally. Well-maintained equipment performs at its best, providing accurate results in labs and efficient operations in workshops.
- Cost Savings: Timely maintenance helps prevent major breakdowns or catastrophic failures, reducing the need for expensive repairs or replacements. Routine maintenance is often more cost-effective than addressing significant issues after they occur.
- Extended Lifespan: Proper maintenance increases the lifespan of equipment, reducing the need for frequent replacements and improving overall cost-effectiveness.

- Compliance: Laboratories and workshops are often subject to regulations and standards that require regular equipment maintenance to ensure compliance with safety and operational guidelines.
- Productivity and Efficiency: Maintained equipment operates more efficiently, leading to increased productivity and reduced downtime.
- Calibration: Regular calibration of laboratory instruments ensures accurate measurements and reliable data, which is critical in scientific research and experiments.
- Environmental Protection: Regular maintenance can prevent leaks, spills, or emissions from equipment, reducing the environmental impact and promoting sustainable practices.
- Reputation: Well-maintained facilities and equipment contribute to a positive reputation for the laboratory or workshop, instilling confidence in customers, partners, and stakeholders.
- Preventive Approach: Regular maintenance takes a proactive approach to identify and address issues before they become significant problems, enhancing overall reliability and safety.
- Asset Management: Proper maintenance helps in managing and tracking assets effectively, ensuring that they are utilized optimally and maintained within budgetary constraints.

In both laboratories and workshops, it is essential to establish a comprehensive maintenance schedule, conduct regular inspections, and document maintenance activities. Maintenance should be carried out by trained personnel or qualified technicians to ensure quality and compliance with industry standards. By prioritizing regular maintenance, laboratories and workshops can create a safer, more efficient, and productive working environment.

b) Routine inspection procedures: Routine inspection procedures for laboratories and workshops are essential to ensure a safe and compliant working environment. Regular inspections help identify potential hazards, equipment malfunctions, and maintenance needs. Here are some general guidelines for conducting routine inspections in labs and workshops:

- Establish Inspection Schedule: Create a regular schedule for inspections based on the specific needs of the laboratory or workshop. Inspections can be conducted weekly, monthly, quarterly, or annually, depending on the nature of the facility and the potential risks involved.

- Inspection Checklists: Develop comprehensive checklists that cover all relevant aspects of safety and compliance. Checklists should include items related to equipment, PPE, emergency systems, storage, hazardous materials, electrical safety, fire safety, and more.
- Equipment Inspection: Inspect all equipment, machinery, and tools for any signs of damage, wear, or malfunction. Ensure that safety features, guards, and interlocks are in place and functioning correctly.
- Electrical Safety: Check electrical systems, outlets, and cords for fraying, damage, or exposed wires. Ensure that electrical components are in good condition and grounded properly.
- Fire Safety: Verify that fire extinguishers are in their designated locations, have valid inspection tags, and are not expired. Inspect fire alarm systems, emergency exits, and sprinkler systems for functionality.
- Chemical Storage: Review the storage and handling of hazardous materials, ensuring proper labeling, segregation, and containment measures are followed.
- Personal Protective Equipment: Check the condition of PPE, such as gloves, lab coats, safety glasses, and respirators. Ensure that employees are using the appropriate PPE for their tasks.
- Emergency Equipment: Inspect eyewash stations, safety showers, and first-aid kits to ensure that they are in proper working order.
- Housekeeping: Assess the overall cleanliness and organization of the laboratory or workshop. Keep workspaces clean and free from clutter to reduce the risk of accidents.
- Training Records: Review employee training records to ensure that all staff have received proper safety training and are knowledgeable about emergency procedures.
- Hazard Identification: Identify any new hazards or changes in the environment that may require additional safety measures.
- Compliance With Regulations: Ensure that the lab or workshop complies with all relevant safety regulations, codes, and standards.

- Corrective Actions: Address any deficiencies or issues identified during the inspection promptly. Assign responsibility for corrective actions and set deadlines for completion.
- Documentation: Maintain detailed records of inspection findings, corrective actions, and follow-up actions taken.
- Review and Improve: Regularly review inspection results and seek opportunities for continuous improvement in safety practices.

By conducting routine inspections and promptly addressing any safety concerns, labs and workshops can maintain a safe working environment and prevent accidents or incidents that may lead to injuries or property damage.

c) Record keeping and documentation: Record-keeping and documentation are essential aspects of maintaining a safe and organized lab and workshop environment. Proper documentation helps track activities, ensure compliance with regulations, and facilitate efficient operations. Here are some key aspects of record-keeping and documentation in labs and workshops:

- Inventory Records: Keeping detailed records of all equipment, materials, and chemicals in the lab or workshop. This includes information such as item names, quantities, dates of acquisition, and expiration dates (if applicable).
- Safety Data Sheets (SDS): Maintaining a comprehensive collection of SDSs for all hazardous chemicals used in the lab or workshop. SDS provide crucial information on handling, storage, and emergency response measures for each chemical.
- Equipment Maintenance Logs: Keeping logs of equipment maintenance and calibration. This includes details of maintenance performed, dates, and signatures of responsible individuals.
- Incident Reports: Recording any accidents, injuries, or near-miss incidents that occur in the lab or workshop. Incident reports should include a description of the event, date, time, location, individuals involved, and any corrective actions taken.
- Training Records: Maintaining records of employee or student training sessions related to lab safety, equipment usage,

and other relevant topics. This ensures that everyone in the lab or workshop has received proper training.

- Experiment Logs and Notebooks: For research labs, maintaining detailed experiment logs and notebooks is critical. These records document experimental procedures, observations, and results, enabling reproducibility and aiding in research integrity.
- Waste Disposal Records: Keeping records of hazardous waste disposal, including the types and quantities of waste, disposal methods, and relevant regulations compliance.
- Permitting and Licensing Documentation: Ensuring all necessary permits, licenses, and certifications are up to date and appropriately documented.
- Inspection and Audit Reports: Keeping records of internal and external inspections, audits, and safety assessments conducted in the lab or workshop.
- Standard Operating Procedures (SOPs): Maintaining a library of Standard Operating Procedures that detail specific procedures, safety protocols, and best practices for various activities in the lab or workshop.
- Project Documentation: In workshops, documenting project plans, progress, and outcomes is crucial for quality control, client communication, and future reference.

Having well-organized and comprehensive records in both labs and workshops promotes accountability, safety, and regulatory compliance. Additionally, it helps in troubleshooting, continuous improvement, and knowledge transfer when new personnel join the lab or workshop. Regular review and updating of records are essential to ensure accuracy and relevance.

It is important to note that regular maintenance and inspections should be conducted by qualified personnel trained to identify potential hazards and make necessary repairs or replacements. Additionally, maintenance and inspections should be conducted per manufacturer specifications and applicable regulations or guidelines.

Regular training and communication among lab personnel are also critical to ensuring that everyone knows the importance of regular maintenance and inspections and is committed to maintaining a safe working environment.

References

1. Al-Zyoud, W., Qunies, A.M., Walters, A.U., Jalsa, N.K., Perceptions of chemical safety in laboratories. *Safety*, 5, 2, 21, 2019.
2. Richards-Babb, M., Bishoff, J., Carver, J.S., Fisher, K., Robertson-Honecker, J., Keeping it safe: Chemical safety in the high school laboratory. *J. Chem. Health Saf.*, 17, 1, 6–14, 2010.
3. Miller, K.A. and Tyler, K.I., Impact of a pilot laboratory safety team workshop. *J. Chem. Health Saf.*, 26, 3, 20–26, 2019.
4. Czornyj, E., Newcomer, D., Schroeder, I., Wayne, N.L., Merlic, C.A., Proceedings of the 2016 Workshop Safety By Design–Improving safety in research laboratories. *J. Chem. Health Saf.*, 25, 4, 36–49, 2018.
5. Adisoemarta, P.S., Riyanto, H., Mustar, M.M., Adrianto, A., Sharing Safety and Safe behavior to Universities thru Lab Safety Workshops, in: *SPE Asia Pacific Oil and Gas Conference and Exhibition*, SPE-147820, 2011.
6. Bharier, J., Managing safety in the laboratory and workshop, in: *Safety in Museums and Galleries*, pp. 93–98, Butterworth-Heinemann, Great Britain, 1987.
7. Gibson, J.H. and Wayne, N.L., Proceedings of the 2012 University of California Center for Laboratory Safety Workshop. *J. Chem. Health Saf.*, 20, 1, 4–17, 2013.
8. Zubir, B., Saad, F., Maidin, W.F., Jamaludin, M.R., Safety and health awareness among staff and students in workshop and laboratory of an Engineering Technology University Campus. *Int. J. Eng. Res. Technol. (IJERT)*, 5, 3, 285–288, 2016.

Section 2
INDUSTRIAL SAFETY

9

Introduction to Industrial Safety Engineering

Abstract
This chapter deals with the basics of safety and its necessity in an organization, types of unsafe acts, and working conditions. Different stakeholders who are responsible for safety in an organization and different accident causation models are also discussed in detail. Lastly, hazard theory is discussed in detail, followed by the hazard triangle, hazard recognition, different basic types of risk, and risk assessment, which have also been described. Also, prevention through design was discussed, which can help in detecting the hazards or risks at an early stage, which can help in avoiding any accidents.

***Keywords*:** Safety engineering, industrial safety, accident causation model, hazard theory, hazard triangle, risk assessment, prevention through design

9.1 Background

Safety is critical in any business because an accident-free work atmosphere boosts the morale of team members who labor in hazardous circumstances. Considering these facts, sectors involving different dangers and risks create their own safety policies and guides, as well as a distinct safety department or division, in order to raise safety consciousness and provide safety know-how. Adherence to helpful information, rules, and statutory requirements regulating safety and standards will aid in the prevention of occupational injuries and mishaps, which are an avoidable and unnecessary loss of human and material resources.

9.2 Introduction

Industrial safety is concerned with the areas of safety engineering and management of all activities and events within an industry that are concerned

Abhishek Barua, Kanchan Kumari, Sumit Saha, Siddharth Jeet and Swastik Pradhan. Institutional and Industrial Safety Engineering Practices, (181–198) © 2025 Scrivener Publishing LLC

with the preservation of employees' health through control of the work environment to reduce or eradicate hazards, risks, accidents, and close misses. Industrial accidents and hazardous working circumstances can cause transient or lasting injury, sickness, or even mortality. They also have an effect on decreased efficiency and output. Industrial safety is governed by federal, state, and municipal rules and regulations. Work-related accidents can be caused by a variety of exterior variables such as chemical, biological, or physical dangers.

One-fifth of the production time is lost by those who are truly hurt due to transient and chronic disablement, and the remainder is lost by fellow workers or people who assist the injured, take care of the damage caused by mishaps, and so on. When death instances from accidents are examined, the loss to the business entity appears much more concerning. It is therefore critical to recognize and investigate the reasons of workplace accidents in order to control them. This safety strategy concerns many fields. One area that works with the creation of efficient work environments, tools, and industrial layouts is industrial engineering. Psychology, sociology, and medicare science are some other fields that can help to create a safe workplace atmosphere. In recent years, engineers have tried to create a systems strategy to workplace accident prevention (dubbed "safety engineering"). The systems method investigates all job areas to identify and eliminate hazards. It also investigates operational methods and practices, as well as employee and management instruction. To successfully and efficiently eliminate an unsafe working atmosphere, adopt the following steps: Elimination of mishap reasons, if feasible, and if it is not feasible to remove the source of the mishap, then make plans to protect the dangerous area with guardians, fences, or other similar arrangements.

9.3 Safety Engineering

Safety engineering is a branch of engineering concerned with the avoidance of accidents, the reduction of dangers connected with human error, and the derivation of safety advantages from engineered systems and designs. It is related to industrial engineering and system engineering and is used in public works, production, and product design to incorporate safety into processes. The goal of safety engineering is to manage danger by reducing or eliminating it entirely. It also seeks to decrease the number of failures, and if a failure does occur, then it is not fatal. Typically, safety engineering starts during the creation of a system or product.

9.4 Need for Safety

Given the aforementioned, the need for and worry for safety is critical. A mishap has some direct costs and effects, but it also has some secondary costs, such as machine delay, machine damage, optimal scheduling of adjacent equipment, worker terror, and loss of time, in addition to settlement costs, legal consequences, and associated costs, etc. As a result, safety measures would suggest not only eliminating or avoiding high-cost risks but also meeting their moral obligation to employees and operators. An accident happens when one or more unsafe factors join to create a hazardous situation. An unsafe act occurs when the operator or others act without appropriate power, misuse safety devices, disregard alerts and safeguards, and so on. A hazardous situation can appear in a variety of ways, such as flawed or defective electrical connections or insufficient gangway upkeep use of faulty equipment, etc. To avoid the incidence of accidents, unsafe actions must be prevented, eradicated, or controlled.

9.5 Types of Unsafe Acts

Attention must be given to the following elements in order to correct the reasons of unsafe acts:

(1) Adjustment of Personnel: If a foreman or superintendent determines that a worker is physically or psychologically unfit for a job or task, then he should be removed from the job immediately in conversation with the personnel department.

(2) Use of Technique: Some tactics that require modification should be substituted with safe alternatives.

(3) Training of Operator: Whether the work technique is secure or unsafe, the operator must be taught to do the job.

(4) Accident avoidance publicity and education: Workers are guided by the ability, energy, and guidance of their foreman or supervisor. As a result, it is their responsibility to teach employees on accident prevention. The aim is to educate them to be safety aware so that they can identify an unsafe act or circumstance and respond appropriately to prevent accidents.

9.6 Unsafe Working Conditions

Various clauses in the "Factory Act" have been debated in order to prevent accidents caused by unsafe circumstances. (Each nation has its own factory act, which is governed by its ruling organizations.) These may be concerned with machine moving components, electrical equipment, fire prevention systems, and so on. Thus, industrial safety is beneficial:

(i) Increasing the pace of output.
(ii) Lowering manufacturing costs.
(iii) Reducing apparatus and equipment harm.
(iv) Preventing unnecessary suffering and agony for the organization's workers.
(v) Preventing the unexpected or premature mortality of gifted employees who could be a benefit to the business and society.

9.7 Safety Programs

A safety program seeks to determine when, where, and why mishaps occur. A safety initiative, on the other hand, seeks to reduce accidents and related costs. A safety program is started with the premise that most workplace mishaps can be avoided. A safety program is a constant process that attempts to reduce the impact of personal and external variables that cause mishaps. A safety program typically consists of giving workers or employees with safety equipment and specific instruction. The Indian Standards Institute has done an excellent work in this regard, as stated below:

(i) Preventive measures to be implemented during manifesting activities.
(ii) Requirements for appropriate lighting, ventilation, and manufacturing unit architecture.
(iii) Standards and specifications for secure manufacturing processes and practices, for example.
(iv) Requirements for efficient tool and machine repair.
(v) Instructions for secure cutting and welding operations.
(vi) Instructions for the use of powered industrial vehicles, belt conveyors, and fire suppression apparatus.
(vii) Personal protective apparatus safety standards.

(viii) Hazardous chemical classification and providing of mishap provision tags.
(ix) Handling and labeling markings for dangerous objects and commodities.
(x) Safety standards: (a) in an industrial structure; (b) in electrical work; and (c) in the use of electrical equipment in hazardous areas and explosive atmospheres.
(xi) Specifications for protective garments, safety helmets, facial barriers, and eye, ear, limb, hand, and foot protection, among other things.
(xii) Concerns about safety and health.

Several employees in various sectors have a work-related injury or sickness. Businesses bear a high cost of workplace accidents and illnesses, with sick days slowing output and penalties levied for bad health and safety procedures. Every workplace will have its own set of health and safety challenges to overcome, but there are some problems that every company, big or small, will be acquainted with.

The impact of any accident leads to health and safety issues. Figure 9.1 shows the plot of time of impact vs. severity of impact. It is health issues if the impact is long term. If immediate or short term, then it is usually safety-related. There is an overlap between health and safety. The overlapping phase is illustrated by the following example: a worker exposed to a material handling job will have a prolonged posture, which will result in

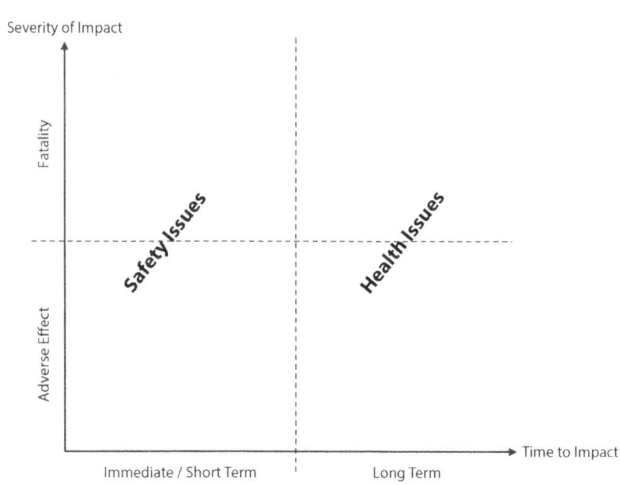

Figure 9.1 Plot of safety and health issues.

musculoskeletal disorder and other health problems. However, when performing material handling, the same worker may be struck by an object or have something fall on him, posing an immediate safety risk. By investing in exceptional health and safety training, the number of injuries will reduce, as will the risk of litigation for any business.

9.8 Stakeholders

Figure 9.2 shows the stakeholders who are responsible for the safety of an organization, i.e., regulatory authorities, economy, organizational/managerial, operational/maintenance, technical/design, research, and analysis. Regulatory authorities are a set of governmental acts designed to ensure safety in any industry. A design engineer who has designed the facility, the product, the machine, and the process is responsible for safety problems and also for failures. The operational maintenance people are responsible for monitoring safety devices and other machinery, which ensures accident prevention. Organizational management is contributing in terms of accident investigation, safety inspection, and regulators. Economics ensures insurance for managers, complex technician operators, worker engineers, system designers, etc.

9.9 Accident Causation Model

It is a methodical technique for determining the reasons behind the accident, which is a complicated coincidence of events or occurrences that happens at the same time and place. As a result, determination of the causes of any accident could be challenging due to large number of factors to study. Accident cause models range from simple linear to complicated non-linear models.

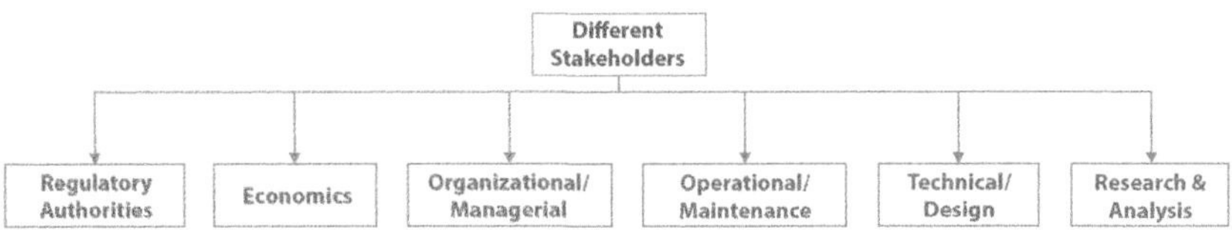

Figure 9.2 Different stakeholders in safety of an organization.

9.9.1 Simple Sequential Linear Accident Models

Accidents, according to basic linear models, are the outcome of a string of events or conditions that relate progressively to one another in an undeviating manner and can be avoided by removing any of the sequential causes.

9.9.1.1 Domino Theory

The domino theory, introduced by safety expert H. W. Heinrich in 1931, was the first sequential mishap model. The model assumes that the incident of an avoidable injury is the natural conclusion of a sequence of events that inevitably arise in a set or rational order. This model suggested that certain mishap variables can be visualize like falling dominoes. The goal of mishap avoidance is in the center of the sequence. It can be visualized as when one accident cause is streamlined with another cause the chances of mishap happening increases just like if domino fells on the other dominoes in series, the whole arrangement collapsed. It is also regarded as knockdown effect (Figure 9.3). Using this model, accident prevention can be done by eliminating any one of the causing factors to halt the knockdown effect.

9.9.1.2 Loss Causation Model

George L. Germain and Frank E. Bird maintained the progressive domino demonstration, acknowledging that Heinrich's domino sequence had supported safety reasoning for three decades. Their model depicted the direct link between management with causes and effects of accident loss. This model was labeled as the loss causation model (Figure 9.4).

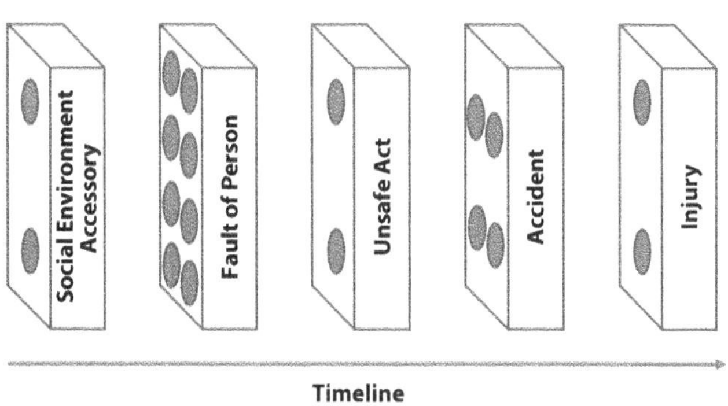

Figure 9.3 Accident causation model of domino theory.

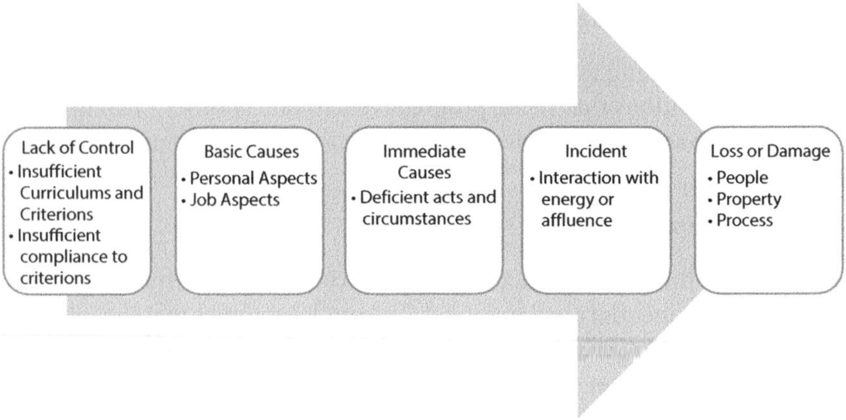

Figure 9.4 Loss causation model.

9.9.2 Complex Linear Models

Accidents are assumed to be the outcome of a mix of risky actions and hidden risk environments surrounded by the system that follow a linear route in complex linear models. The factors farthest away from the accident are attributed to organizational or environmental actions, whereas the variables at the severe end are where humans eventually engage nearest to the accident; the implication is that accidents prevention can be done by concentrating on building obstacles and defenses.

9.9.2.1 Energy-Damage Model

The energy damage model (Figure 9.5) is based on Derek Viner's 1991 proposal that harm is initiated by instance energy whose strength at the site of interaction with the receiver surpasses the recipient's damage threshold. In this model, the hazard is a potentially harmful energy source, and an

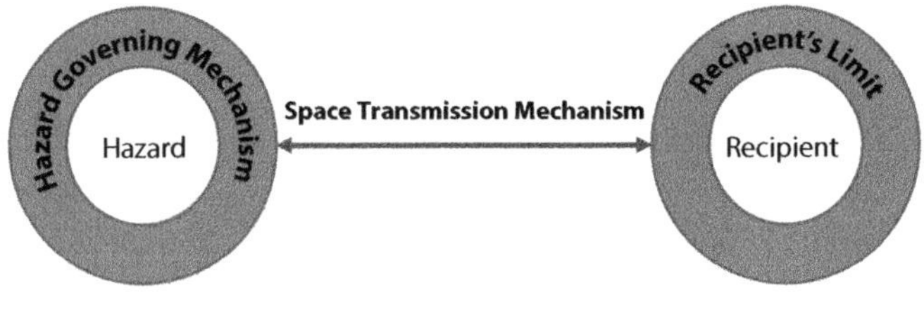

Figure 9.5 Energy damage model.

injury, mishap, or damage may occur due to a loss of energy control when the mechanism of hazard control fails. These systems include structural or physical containment, techniques, obstacles, and routines. The space transmission process is the technique by which recipient and energy, assuming that they are initially divided, are taken together. The exposed and susceptible surface to energy is the receiver limit.

9.9.2.2 *Time Sequence Models*

Figure 9.6 shows Derek Viner's time-sequence model. Viner thinks that the framework for analyzing events in the occurrence-consequence sequence provided by the time sequence model draws attention to defenses that would otherwise be ignored. In Time Zone 1, there is a possibility that the event can be avoided. Time Zone 2 warns the imminent presence of an event mechanism as well as the chance to take measures to which can reduce the chance of the event to occur. Time Zone 3 provides a prospect to impact the consequence and the vulnerable groups.

9.9.2.3 *Epidemiological Model*

Approaches of accident prevention that correlate to an epidemiological accident model emphasize reducing errors and understanding the accident's hidden causes. Such reasons can be noted in violations or dangerous actions, and suppressing or removing them can help avoid a recurrence of the mishap. Moreover, thinking about safety protection is moving toward reality that processes must be strong enough to withstand abnormalities or odd actions without possible disruption.

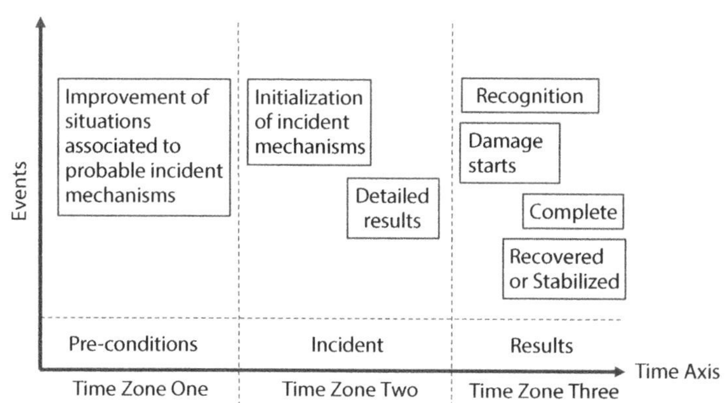

Figure 9.6 Time sequence model.

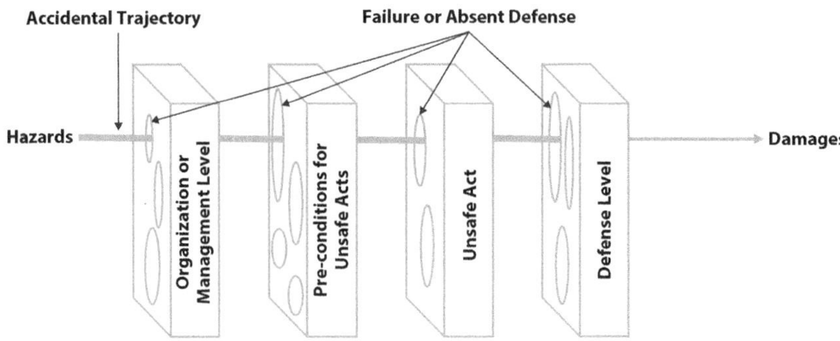

Figure 9.7 The Swiss cheese model.

9.9.2.4 *Swiss Cheese Model*

James Reason introduced Reason's model, alias the Swiss Cheese Model, in 1990 (Figure 9.7). He created representations of human error processes by exploring daily slips and lapses and tackled the problem of active and latent errors. In active errors, the impact is felt almost instantaneously, whereas latent errors tended to lay inert in the system largely unnoticed until they joined with other variables to penetrate system defenses. The explanation provided was that accidents were caused by broader systemic organizational variables at the higher tiers of the organization, rather than individual operator error.

9.9.3 Complex Nonlinear Accident Models

The accident models tend to be non-linear in latest trends, i.e., accidents can be predicted from commonly relating variables that arise in real-world environments. In the early 2000s, two novel models were developed to solve linear accident models problems.

9.9.3.1 *Systems-Theoretic Accident Model and Process (STAMP)*

STAMP model was formulated by Dr. Nancy Leveson at MIT, which is based on systems theory thinking. STAMP incorporates into industrial investigation causative factors like software, new technology, human factors and human decision-making, safety culture, and organizational and social design, all of which are increasingly dangerous for complicated systems.

9.9.3.2 *Functional Resonance Accident Model (FRAM)*

FRAM model was formulated by Erik Hollnagel, and it offers a method to characterize conclusions using the concept of resonance resulting from the inconsistency of daily enactment. A FRAM analysis consists of the following stages:

- Identification, explanation, and characterization of critical system functions using the six fundamental traits (aspects). Only describe the elements that are essential or pertinent in the first version. It is always possible to change the summary afterward.
- Examine the model's thoroughness and coherence.
- Identify the prospective inconsistency of the FRAM model's functions.
- Outline functional resonance centered on function relationships and correlations, as well as the possibility for useful inconsistency.
- Identification of methods for monitoring the evolution of resonance.

9.10 Hazard Theory

The theory of hazards in terms of practical application forms the basis for safety engineering as a separate discipline of technical sciences. The theory of hazards arises from the unification of the science of objects' reliability and unsafety. So, a hazard is a potentially dangerous situation that existed before the accident (Figure 9.8). This hazard is converted into an accident, and that is known as the state condition. So, there is a state of transition in any system, workplace, or job. Hazards are unintentionally constructed into a system and can be identified by its components. A design defect can be an accident waiting to happen; thus, a safety engineer's task in an

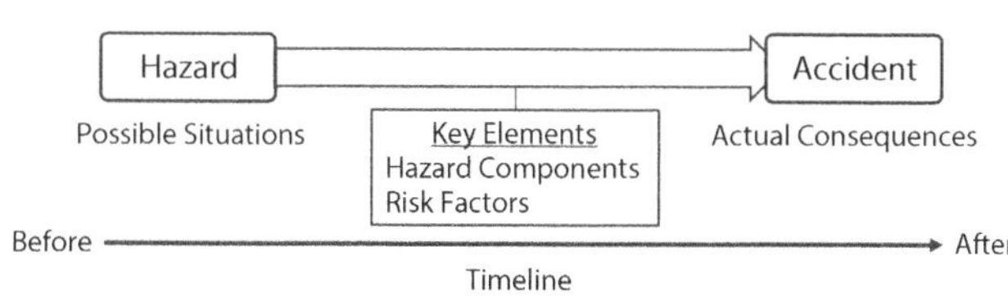

Figure 9.8 Basic hazard theory.

organization is to understand how this possible situation ultimately leads to an accident and to know the route. A hazard is both a deterministic object and a random occurrence that occurs as a result of the hazard's components. Accidents and hazards are foreseeable and, thus, avoidable and controllable.

9.11 Hazard Triangle

The Hazard Triangle, also known as the "safety triangle," provides normal distributions and correlation of close misses, minor accidents, and serious or deadly accidents (Figure 9.9). It exemplifies the significance of a comprehensive strategy to crisis management.

- Hazardous element (HE): They are fundamental hazardous substances that generate the hazard, e.g., the use of explosives in the system. High-pressure containers, fuel, and high voltage are a few instances.
- Initiating mechanism (IM): This triggers or initiates the danger. The initiating process causes the danger to become active or to be activated from a dormant condition. A few instances include a tank rupture, a gasoline spill, an ignition source, contacting an open contact, and so on.
- Target and danger (T/T): This is the object or any person that is at risk of harm or damage, as well as the seriousness of the accident occurrence. This is due to the accident, as well as the anticipated harm and loss. Explosions, fires, and electrocution are a few instances of events that can result in mortality or harm.

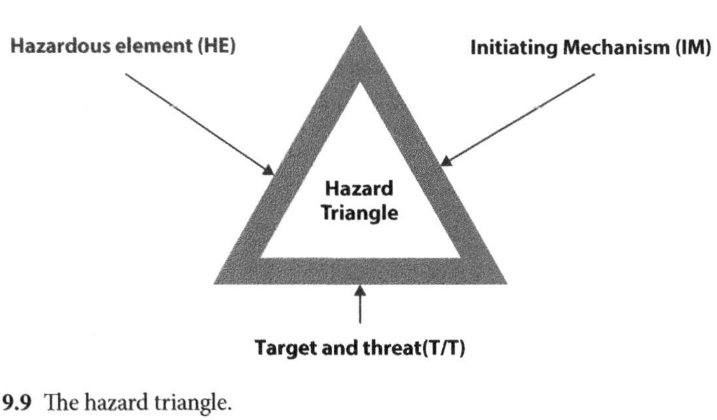

Figure 9.9 The hazard triangle.

9.12 Hazard Recognition

The hazard recognition process includes an understanding of hazard theory, which will provide a consistent and methodical process to deal with any mishaps. It also includes an understanding of system design and operation. There are many key recognition factors, such as

- Utilization of the hazard triangle components: use of the HE/component checklist (HE), evaluation of trigger events and causal factors (IM), and evaluation of possible threats and mishaps [target and threat (T/T)].
- Applying prior experience and lessons gained, as well as analyzing sound design practices.
- A study of general design safety criteria, concepts, and precepts is performed.
- The evaluation of top-level mishaps and safety-critical tasks, as well as the key failure condition issue.

9.13 Individual Risk and Societal Risk

9.13.1 Individual Risk (IR)

Individual risk is the likelihood where an individual within a specific group or community will suffer a given degree of harm as a result of the occurrence of a specific hazard or specific combinations of hazards. It can be computed as follows:

$$Individual\ Risk = \frac{Societal\ Risk}{Population\ at\ Risk} \tag{19.1}$$

Individual risk per annum (IRPA) is described as the number of people slain in a single year of exposure. It can be calculated as a safety efficiency indicator as

$$IPRA = \frac{Total\ no.\ of\ fatalities}{Total\ no.\ of\ employee\ in\ years\ exposed} \tag{19.2}$$

The individual risk index is the likelihood that an unprotected person continuously exist at a specific place will be killed in a year due to a mishap caused by a hazardous activity. It is primarily employed in land-use planning. It can be computed as follows:

$$Individual\ Risk\ Index(x,y) = \Sigma_{i=1}^{m}\lambda_i \times Fatality\ at(x,y)|A_i \quad (19.3)$$

where A_i signifies type "i" accident, and λ_i denotes frequency of A_i.

9.13.2 Societal Risk (SR)

The frequency and quantity of people who experience a particular degree of harm as a result of the occurrence of defined hazards is referred to as societal risk. It can be computed as follows:

$$Societal\ risk = Frequency \times Magnitude \quad (19.4)$$

where Societal risk is consequences per time, Frequency is events per time, and Magnitude is consequences per event.

9.14 Risk Assessment

A risk assessment is an all-inclusive examination of a workplace finding those objects, procedures, and circumstances, which causes damage (Figure 9.10) and then removes or reduces their degree of risk by implementation of control measures as per requirement. When new developments or activities are presented into established processes or activities, or when hazards are discovered, risk evaluation should be performed. In general, when conducting an assessment, one should:

- Detect probable risks.
- Determination of chance and degree of damage, for instance, an accident or sickness.
- Determination of the steps required for the elimination of danger or managing the risk with the help of risk control hierarchy.

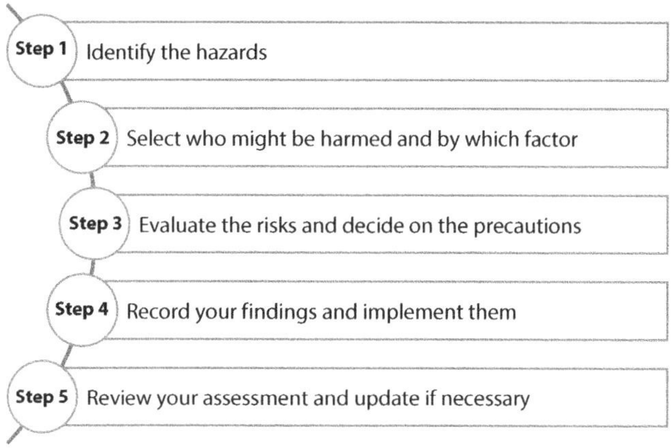

Figure 9.10 Basic steps of risk assessment.

- Evaluation for ensuring the removal of danger or proper management of the risk.
- Monitoring the situation for ensuring the effectiveness of the management.
- Maintaining the records or documents, which include description of the method used for the evaluation of risk, an outline of any assessments, and an explanation of how findings were reached.

It can be noted that assessment should consider both present condition of the workplace along with any possible situations. The safety management of the organization can determine the extension of control program (if needed) by finding the degree of risk associated with the danger.

9.14.1 Quantitative vs. Qualitative

Risk evaluations can be either objective or intuitive. In a quantitative risk evaluation, the organization's risk manager gives numerical values to the likelihood of an incident occurring and the effect it would have. These number values can then be used to determine an event's risk factor, which can then be converted to a monetary value. The more common qualitative risk evaluations do not include numerical odds or loss forecasts. A qualitative method simply ranks which risks offer the greatest hazard.

9.14.2 Risk Prioritization Using Matrix

Ranking risks is used for determination of severe risk among all which should be addressed first. Priority is usually decided by the employee's exposure and the chance of an accident, injury, or illness. Understanding workplace activities, the urgency of circumstances, and, most significantly, impartial assessment are all required for ranking dangers. Checklists or a likelihood matrix can be useful in some situations. A team of knowledgeable personnel who are acquainted with the job is generally required for more complicated circumstances. Figure 9.11 shows a typical risk assessment matrix where the priority of a risk is decided based on the probability and severity of an accident occurring.

9.14.3 Reviewing Risk Assessment

It is critical to understand whether your risk evaluation was thorough and correct. It is also critical to ensure that any workplace changes have not inducted new dangers or shifted hazards that were previously listed as lower priorities to higher priorities. It is recommended for assessment evaluation on a frequent basis for ensuring control techniques are successful. It is critical to keep documents of your evaluation and any control actions done. You might be obliged to keep evaluations for a certain number of years. Check your jurisdiction's municipal regulations.

Probability	Severity			
	Catastrophic (1)	Critical (2)	Marginal (3)	Negligible (4)
Frequent (A)	High	High	Serious	Medium
Probable (B)	High	High	Serious	Medium
Occasional (C)	High	Serious	Medium	Low
Remote (D)	Serious	Medium	Medium	Low
Improbable (E)	Medium	Medium	Medium	Low
Eliminated (F)	Eliminated			

Figure 9.11 Sample of risk assessment matrix.

9.15 Prevention Through Design

One of the most operative methods for prevention and controlling of workplace accidents, illnesses, and deaths is to design out or reduce dangers along with risks. Prevention through design, also known as "safety by design," refers to all determinations to foresee and eliminate dangers to employees in buildings, processes, operations and work methods, equipment, products, new technologies, tools, and work organization. Prevention through design focuses on employees who carry out the designs or deal with the products of the designs. The program was created to assist the most dependable and efficient form of prevention: designing out risks (Figure 9.12). This can be accomplished by the following:

- Reducing risks and hazards to employees to an acceptable limit at the beginning in the life cycle of workplaces or products as feasible.
- Designing, restructuring, and retrofitting current and new work places, structures, facilities, tools, equipment, machines, work processes, goods, substances, and organizational structures.
- Enhancing the work atmosphere by incorporation of preventive methods in all designs that affect employees and others present on the grounds.

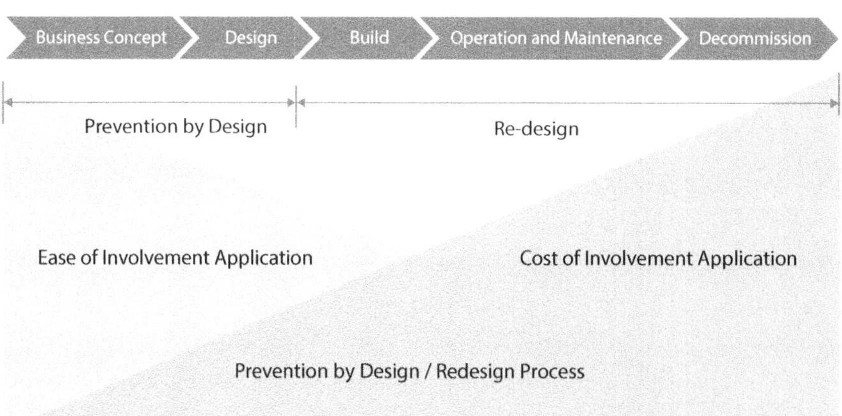

Figure 9.12 Prevention through design chart.

The tactical plan sets objectives for effective prevention through design execution. The plan, which incorporates worker's safety and health into all parts of design, will serve as a critical foundation for saving lives, avoiding occupational accidents and illnesses.

Bibliography

Calavita, K., The demise of the Occupational Safety and Health Administration: A case study in symbolic action. *Social Probl.*, 30, 4, 437–448, 1983.

Kogi, K., Work improvement and occupational safety and health management systems: common features and research needs. *Ind. Health*, 40, 2, 121–133, 2002.

Maiti, J. and Ray, P.K., *Industrial Safety Management*, Springer, Singapore, 2018.

Ray, P.K. and Maiti, J., *Ergonomic Design of Products and Worksystems-21st Century Perspectives of Asia*, Springer, Singapore, 2017.

Toft, Y., Dell, G., Klockner, K., Hutton, A., Models of causation: Safety, Safety Institute of Australia, 2012.

10

Hazard Identification and Analysis Techniques

Abstract
This chapter describes hazard identification in any organization, followed by its importance and important steps to conduct it. In the latter part, different hazard identification and analysis techniques are discussed, with a detailed discussion provided about their significance, important steps to conduct analysis, and their advantages and disadvantages.

Keywords: Hazard identification, hazard analysis, HAZOP, fault tree analysis, preliminary hazard analysis, bow tie analysis, FMEA, job hazard analysis

10.1 Background

The origins of hazard identification and analysis methodologies may be traced back to the growth of industrial and technical processes, as well as the growing awareness of the importance of safety and risk management. Today, hazard detection and analysis approaches are evolving in response to new technology, sectors, and regulatory mandates. They are critical components of risk management and safety programs in a variety of industries, such as manufacturing, healthcare, transportation, and environmental protection. These strategies assist organizations in proactively identifying and mitigating possible risks in order to avoid accidents, protect workers and the public, and guarantee the safe functioning of processes and systems.

10.2 Introduction

Hazard identification is a critical stage in managing risks and safety. It entails detecting and analyzing possible sources of damage, risk, or adverse

Abhishek Barua, Kanchan Kumari, Sumit Saha, Siddharth Jeet and Swastik Pradhan. Institutional and Industrial Safety Engineering Practices, (199–266) © 2025 Scrivener Publishing LLC

occurrences within a particular environment, system, process, or activity. The ultimate purpose of hazard identification is to recognize and comprehend possible dangers so that suitable preventive or mitigating actions can be implemented. This process is critical in a wide range of businesses and situations, including manufacturing and construction, as well as healthcare, transportation, and environmental protection. It enables organizations to identify possible threats, take preventative measures, and create safer settings for employees, the general public, and the environment [1].

Techniques for identifying and analyzing hazards are critical procedures in many sectors for assessing and mitigating risks connected with operations, goods, or projects. These approaches assist organizations in identifying possible dangers, assessing their effects, and developing risk-mitigation strategies. The three primary risk concerns are as follows:

- Danger: What might go wrong?
- Effects: How bad might it get?
- Probability: How likely is it that it will occur?

While addressing these questions, the goal is to undertake only the degree of investigation required to make a choice, because inadequate analysis may result in incorrect choices and unnecessary research wastes resources. It is worth noting that when risk assessments grow more concentrated and precise, the cost of each scenario analyzed rises, but the total expenditure may fall if only a few typical or boundary situations are examined.

10.3 Importance of Hazard Identification

The importance of hazard identification cannot be overstated, as it plays a critical role in ensuring safety, preventing accidents, protecting the environment, and maintaining the overall well-being of individuals and organizations. Hazard identification is essential for minimizing risks, protecting people and the environment, complying with regulations, and ensuring the long-term sustainability of organizations [2]. It is a cornerstone of effective risk management and safety culture, and its importance extends across various industries and sectors. Here are several key reasons why hazard identification is of utmost importance:

- Accident prevention: Hazard identification is a proactive approach to accident prevention. By identifying potential hazards before they lead to accidents or incidents, organizations can take preventive measures to reduce the risk of harm to employees, the public, and property.

- Worker safety: In workplace environments, hazard identification is essential for protecting the health and safety of employees. It helps in recognizing and mitigating risks associated with machinery, equipment, chemicals, ergonomics, and other factors that can cause injuries or illnesses.
- Environmental protection: Hazard identification extends beyond the workplace to include potential risks to the environment. Recognizing and addressing environmental hazards is vital for minimizing pollution, ecological damage, and harm to ecosystems.
- Legal compliance: Many jurisdictions have regulations and standards that require organizations to identify and mitigate hazards. Compliance with these legal requirements is essential to avoid penalties, fines, and legal liabilities.
- Financial consequences: Accidents and incidents resulting from unidentified hazards can have significant financial consequences. These may include medical expenses, property damage, legal fees, increased insurance premiums, and loss of productivity. Hazard identification helps mitigate these financial risks.
- Operational continuity: Hazard identification is crucial for ensuring the uninterrupted operation of systems and processes. By addressing potential risks, organizations can prevent disruptions that could impact their productivity and profitability.
- Public safety: Hazard identification is not limited to the workplace; it also applies to public spaces, transportation systems, and infrastructure. Ensuring the safety of the public is a fundamental responsibility, and hazard identification is a key component of achieving this goal.
- Emergency preparedness: Identifying hazards is essential for emergency preparedness and response planning. Knowing the potential risks allows organizations to develop effective strategies for responding to emergencies and minimizing their impact.
- Continuous improvement: Hazard identification is an ongoing process that promotes continuous improvement. Regularly assessing and addressing hazards leads to a safer

work environment, improved processes, and reduced risks over time.

- Reputation and stakeholder trust: Organizations that prioritize hazard identification and safety build a positive reputation and gain the trust of employees, customers, investors, and regulatory authorities. Conversely, accidents resulting from neglecting hazard identification can tarnish an organization's image.
- Innovation and sustainability: Identifying and addressing hazards can drive innovation in safety technologies and sustainable practices. Organizations that invest in hazard identification often find new and more efficient ways to operate safely and responsibly.

10.4 When is it Done?

Hazard identification and analysis evaluations can take place at any step of a project's life cycle, including initial planning, early design, thorough planning, creation, continuous operation, elimination, and destruction. In general, the sooner a danger is detected (for example, early concept design), the easier it is to eliminate or control it. During the early phases of design, studies are often conducted at commercial or industrial offices. Studies conducted in a plant setting are often conducted while a process is nearing commencement, throughout operation, or before decommissioning.

10.5 Who Prepares Hazard Identification?

Hazard identification is typically a collaborative effort that involves individuals with relevant expertise and knowledge about the specific environment, process, or system being assessed. It is important to note that hazard identification is not a one-time task but an ongoing process. As circumstances change, new hazards may emerge, and existing hazards may evolve. Therefore, organizations should promote a culture of safety where all employees are encouraged to report and address hazards as they are identified. Ultimately, the specific individuals and groups responsible for hazard identification will depend on the nature of the hazards, the industry, and the organization's safety and risk management policies and practices. The responsibility for preparing hazard identification can vary depending on the context, industry, and organization [3]. Here are some common scenarios where hazard identification is prepared and who might be involved:

- Safety professionals: In many organizations, safety professionals or safety engineers play a central role in leading hazard identification efforts. They have expertise in risk assessment and safety management and may use various techniques and tools to identify potential hazards.
- Process engineers: In industries with complex processes, such as chemical manufacturing or oil refining, process engineers often participate in hazard identification. They have a deep understanding of the systems and equipment involved and can identify process-specific hazards.
- Multidisciplinary teams: Hazard identification is often most effective when it involves a multidisciplinary team. This team may include individuals with expertise in engineering, health and safety, environmental science, quality assurance, and other relevant areas. Each team member contributes their knowledge to the hazard identification process.
- Employees and workers: Frontline employees and workers who are directly involved in the daily operations of a workplace are valuable resources for hazard identification. They have firsthand knowledge of the workplace and can identify hazards that may not be immediately apparent to others.
- Regulatory agencies: In some cases, regulatory agencies may conduct hazard identification as part of their oversight and compliance activities. They assess whether organizations are complying with safety and environmental regulations and may identify hazards during inspections.
- Third-party consultants: Some organizations hire third-party consultants or experts with specialized knowledge in hazard identification and risk assessment to conduct assessments independently. These consultants can provide an unbiased perspective on potential hazards.
- Safety committees: In workplaces with established safety committees, these committees may take on the responsibility for hazard identification. These committees typically include representatives from different departments or areas within an organization.

- Community and public input: In cases where potential hazards impact the surrounding community, public input and feedback can be valuable in hazard identification. Public concerns and observations may lead to the identification of hazards that might otherwise be overlooked.

10.6 Some Commonly Used Hazard Identification and Analysis Techniques

10.6.1 Hazard and Operability Study (HAZOP)

HAZOP is a systematic and organized approach used in the areas of safety and risk management, mainly in the chemical, petrochemical, and industrial sectors. A HAZOP study's primary goal is to identify possible hazards and operational concerns related to an upcoming or current process, framework, or facility. A multidisciplinary team of specialists studies the process or system in depth and carefully analyses multiple variables and variations from the planned layout conditions during a HAZOP investigation [4]. Figure 10.1 depicts the HAZOP analysis process.

10.6.1.1 How Does HAZOP Work?

HAZOP is a systematic process safety analysis technique that is frequently utilized in manufacturing sectors where process safety is critical, such as the chemical, pharmaceutical, petrochemical, and oil and gas industries, to identify and assess potential hazards and operability issues within a system, process, or operation. It assists organizations in proactively managing and mitigating risks in order to avoid accidents and maintain safe and efficient operations [5]. Here is a step-by-step overview of how HAZOP works:

(i) Form a multidisciplinary team: Assemble a team of experts from various disciplines, including process engineers, operators, safety professionals, and other relevant stakeholders. This team's diverse expertise ensures a comprehensive analysis.

(ii) Select the node: Define the boundaries of the system, process, or operation that one wants to analyze. This is often referred to as the "node." It could be a specific piece of equipment, a process unit, or a series of steps within a larger process.

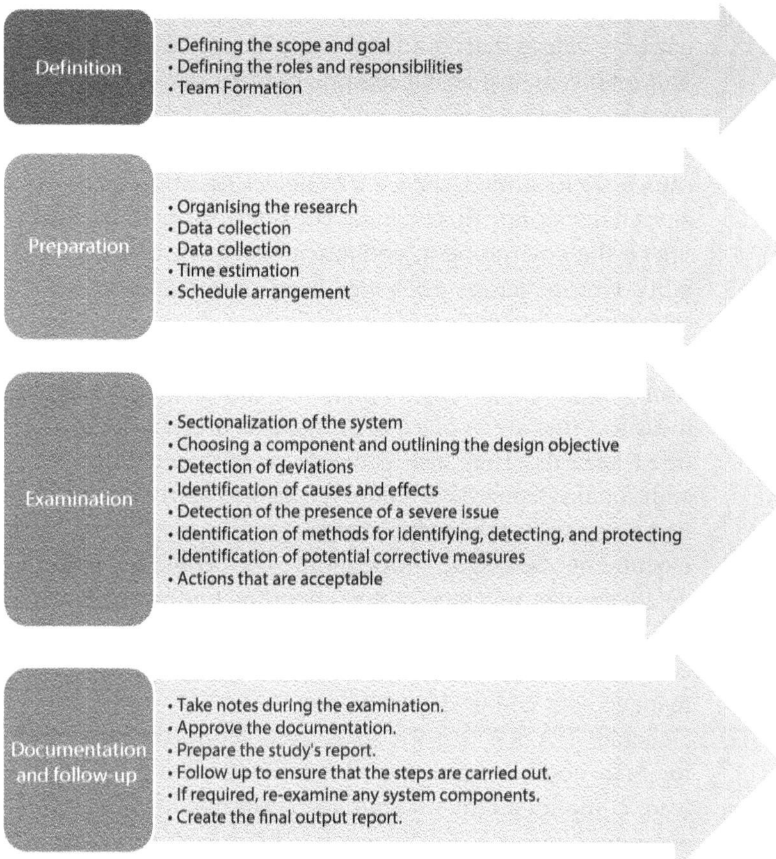

Figure 10.1 The HAZOP analysis process.

(iii) Choose guidewords: The HAZOP team selects a set of guidewords that will be used to systematically explore deviations from the design or operational intent. Common guidewords include "no," "more," "less," "reverse," "early," "late," "high," "low," and "temperature."

(iv) Identify parameters: For each node and selected guideword, the team examines process parameters (e.g., temperature, pressure, flow rate, level, and composition) to identify potential deviations from the intended values. They ask questions like, "What happens if the temperature is higher than expected?" or "What if the flow rate is lower than intended?"

(v) Brainstorm deviations: The team brainstorms various scenarios or causes that could lead to the identified deviations. This involves considering equipment malfunctions, operator errors, external factors, and other potential causes.

(vi) Assess consequences: Once potential deviations and their causes are identified, the team evaluates the consequences of each deviation. This includes considering the impact on safety, the environment, equipment integrity, and operability. Consequences can range from minor nuisances to catastrophic events.

(vii) Identify safeguards and recommendations: The HAZOP team discusses existing safeguards, controls, and safety measures that are in place to mitigate or prevent the identified hazards. They also generate recommendations for additional safeguards or improvements to existing ones, if necessary, to further reduce risks.

(viii) Document findings: Throughout the HAZOP study, detailed notes are taken to document the findings, including deviations, causes, consequences, and recommended actions. This documentation is crucial for subsequent risk management and decision-making.

(ix) Generate HAZOP report: After completing the analysis for each node and guideword combination, the HAZOP team compiles all the findings and recommendations into a comprehensive HAZOP report. This report serves as a formal record of the study's results and provides guidance for risk reduction and process improvement.

(x) Implement recommendations: Based on the HAZOP report, the organization can prioritize and implement the recommended actions to mitigate identified hazards and enhance process safety.

(xi) Review and update: Periodic reviews and updates of the HAZOP study are essential to account for changes in the system, process, or operational conditions over time.

10.6.1.2 Importance of HAZOP

HAZOP is of significant importance to safety professionals and is widely used in industries where process safety is critical. HAZOP is an essential tool for safety professionals because it helps them proactively identify, assess, and manage risks, ultimately leading to safer operations, regulatory compliance, and the prevention of accidents and incidents. It supports their role in safeguarding the well-being of employees, the environment, and the overall success of an organization [6]. Here are several reasons why HAZOP is crucial for safety professionals:

(i) Systematic hazard identification: HAZOP provides a structured and systematic approach to identifying potential hazards within a system or process. Safety professionals can systematically examine various parameters and deviations, helping them uncover risks that might otherwise go unnoticed.
(ii) Risk assessment: HAZOP allows safety professionals to assess the severity and likelihood of identified hazards and their potential consequences. This risk assessment forms the basis for prioritizing actions to mitigate or eliminate risks.
(iii) Early detection of hazards: HAZOP is often conducted during the design or modification phase of a process. Safety professionals can participate in HAZOP studies to identify and address hazards at an early stage, minimizing the risk of accidents during operations.
(iv) Operational safety: HAZOP not only identifies potential hazards but also assesses operability issues. Safety professionals can help ensure that a system or process is not only safe but also efficient and operable.
(v) Regulatory compliance: Many industries have strict safety regulations and standards. Safety professionals can use HAZOP to demonstrate compliance with these regulations and ensure that safety measures are in place to meet legal requirements.
(vi) Incident prevention: By identifying and addressing potential hazards before they lead to accidents, HAZOP helps prevent incidents that could result in injuries, fatalities, environmental damage, and financial losses.
(vii) Continuous improvement: HAZOP is not a one-time exercise. Safety professionals can use it as part of a broader safety management system to continuously improve safety performance. Regular HAZOP reviews help ensure that safety measures remain effective as processes evolve.
(viii) Communication: HAZOP involves multidisciplinary teams, fostering communication and collaboration among

professionals from various backgrounds. Safety professionals play a key role in facilitating these discussions and ensuring that safety concerns are addressed.

(ix) Documentation: Safety professionals are responsible for documenting the findings and recommendations of HAZOP studies. Clear and comprehensive documentation is essential for accountability, decision-making, and future reference.

(x) Training and awareness: Safety professionals can use HAZOP as a tool to raise awareness among employees and operators about potential hazards and safe operating practices. This helps create a safety-conscious culture within an organization.

(xi) Emergency preparedness: Understanding the hazards identified through HAZOP studies enables safety professionals to develop effective emergency response plans and procedures in case of unexpected events.

(xii) Resource allocation: Safety professionals can use the results of HAZOP studies to allocate resources effectively for risk reduction measures. They can prioritize actions based on the severity of identified hazards.

10.6.1.3 Pros and Cons of Conducting a HAZOP Study

HAZOP is a valuable tool for identifying and mitigating risks, but it is not without its challenges. Organizations should consider both the benefits and limitations of HAZOP when deciding to implement it as part of their safety and risk management strategies, but like any method, it comes with its own set of advantages and disadvantages.

Pros of HAZOP:

- Systematic approach: HAZOP provides a structured and systematic approach to identifying potential hazards, making it less likely that critical issues will be overlooked.
- Early hazard identification: HAZOP can be conducted during the design or modification phase of a process, allowing for the early identification and mitigation of hazards, reducing the risk of incidents during operation.
- Multidisciplinary collaboration: HAZOP involves a diverse team of experts from various disciplines, fostering

collaboration and drawing on a wide range of expertise to identify hazards and solutions.

- Operability assessment: HAZOP not only identifies hazards but also assesses operability issues, ensuring that a system or process is both safe and efficient.
- Compliance with regulations: HAZOP helps organizations demonstrate compliance with safety regulations and standards, reducing the risk of legal and regulatory issues.
- Documentation: The process results in detailed documentation of identified hazards, causes, consequences, and recommended actions, serving as a valuable reference for risk management.
- Risk prioritization: HAZOP helps prioritize risks by assessing their severity, likelihood, and detectability, enabling organizations to allocate resources effectively for risk reduction measures.
- Continuous improvement: Regular HAZOP reviews can be integrated into a safety management system, supporting continuous improvement in safety performance.

Cons of HAZOP:

- Resource-intensive: HAZOP studies require a significant investment of time, manpower, and resources, which can be a limitation for smaller organizations with limited resources.
- Complexity: The process can be complex, especially for large and intricate systems, making it challenging to ensure a comprehensive analysis.
- Subjectivity: HAZOP relies on expert judgment and brainstorming, which can introduce subjectivity into the identification of hazards and their consequences.
- Limited focus: HAZOP primarily focuses on process-related hazards and may not address all types of risks, such as those related to organizational factors or human error.
- Resistance to change: Organizations may be resistant to implementing recommended changes or safety measures due to concerns about cost, disruption to operations, or other factors.
- Potential for overlooked hazards: While HAZOP is thorough, it is still possible for some hazards to be missed, especially if they are highly unlikely or unique.

- Difficulty in quantifying risks: HAZOP provides qualitative assessments of risks but may not provide precise quantitative data on risk levels.
- Long-term commitment: Maintaining an effective HAZOP program requires ongoing commitment and resources to ensure that recommendations are implemented and that the process remains up to date.

10.6.1.4 Hazard and Operability Study Tips

Conducting a successful HAZOP requires careful planning, a skilled team, and effective facilitation. HAZOP is not a one-size-fits-all technique, and its success depends on careful planning, effective teamwork, and a commitment to safety. By following these tips and best practices, one can enhance the effectiveness of HAZOP studies and contribute to safer operations in the organization. Here are some tips to help to navigate the HAZOP process effectively:

(i) Select an experienced team: Assemble a multidisciplinary team with relevant expertise in engineering, operations, safety, and other relevant fields. Having a diverse group of participants will enhance the quality of the analysis.

(ii) Prepare adequately: Ensure that all team members are familiar with the system or process under study, its design, and its intended operation. Preparatory meetings or training sessions can help participants understand the HAZOP process and their roles.

(iii) Define clear objectives: Clearly define the scope and objectives of the HAZOP study. What are the specific nodes or areas of the process that need analysis? Setting clear goals helps focus the study.

(iv) Use guidewords effectively: Select and use guidewords appropriately during the analysis. Ensure that team members understand how to apply each guideword to different process parameters.

(v) Keep the sessions focused: HAZOP sessions can be lengthy, so it is essential to stay focused on the task at hand. Avoid distractions and side discussions that do not contribute to hazard identification.

(vi) Document thoroughly: Maintain detailed and accurate documentation throughout the HAZOP sessions. Record deviations, causes, consequences, safeguards, and recommendations systematically. Accurate documentation is vital for the final report and follow-up actions.

(vii) Use visualization tools: Utilize visual aids, such as process flow diagrams or P&ID (piping and instrumentation diagrams), to help participants better understand the system and its components.

(viii) Challenge assumptions: Encourage team members to question assumptions and consider worst-case scenarios. Sometimes, assumptions can lead to the overlooking of potential hazards.

(ix) Prioritize risks: After identifying hazards, prioritize them based on severity, likelihood, and detectability. This prioritization will guide the allocation of resources for risk reduction measures.

(x) Facilitate constructive discussions: The HAZOP facilitator plays a crucial role in maintaining a productive and respectful discussion. Ensure that the facilitator is skilled in managing group dynamics and conflicts, if they arise.

(xi) Consider human factors: Do not forget to consider human factors and human error as potential sources of hazards. Human factors can significantly impact the safety of a system or process.

(xii) Address recommended actions: Once the HAZOP study is complete, ensure that the recommended actions are reviewed, prioritized, and implemented. Follow up on action items to enhance safety.

(xiii) Document and share findings: Prepare a comprehensive HAZOP report summarizing the study's findings, including identified hazards, recommended actions, and risk assessments. Share this report with relevant stakeholders.

(xiv) Continuous improvement: Treat HAZOP as an ongoing process. Regularly review and update the analysis as the system or process evolves over time.

10.6.1.5 The HAZOP Study Process

The HAZOP study process is a rigorous and systematic approach to hazard identification and risk assessment. It involves a series of structured steps that guide a multidisciplinary team through a comprehensive analysis. It helps organizations proactively manage and mitigate risks, ensuring the safety of personnel, the environment, and the integrity of operations. Here is an overview of the HAZOP study process:

1) Preparation and planning:
    - Define the scope and objectives of the HAZOP study.
    - Select and assemble a team of experts with diverse backgrounds and relevant expertise.
    - Identify the system, process, or operation to be analyzed (the "node").
    - Determine the process parameters and guidewords that will be used during the study.
    - Establish a timeline and schedule for the HAZOP sessions.
2) Node selection:
    - Clearly define the boundaries of the system or process to be analyzed. This is often referred to as the "node."
    - Create or gather the necessary documentation, such as process flow diagrams, P&IDs, and equipment specifications, to provide a clear understanding of the node.
3) Guideword application:
    - Select and apply guidewords systematically to different process parameters within the node. Guidewords represent potential deviations from the intended design or operational parameters.
    - For each guideword, consider how it could affect the process parameters and lead to deviations or hazards.
4) Deviation identification:
    - Brainstorm and identify potential deviations or scenarios for each combination of guideword and process parameter. Ask questions like, "What happens if the temperature is higher than expected?" or "What if the flow rate is lower than intended?"
    - Document these deviations along with their causes, if known.
5) Consequence assessment:
    - Evaluate the consequences of each identified deviation, considering safety, environmental impact, equipment integrity, and operability.
    - Consequences can range from minor issues to major hazards.

6) Safeguard analysis:
   - Discuss and identify existing safeguards, controls, and safety measures in place to mitigate or prevent the identified hazards.
   - Generate recommendations for new safeguards or improvements to existing ones if necessary.
7) Documentation:
   - Maintain detailed records of the HAZOP study findings, including deviations, causes, consequences, and recommended actions.
   - Accurate documentation is essential for the final report.
8) Report preparation:
   - Compile all the findings and recommendations into a comprehensive HAZOP report.
   - The report should include a summary of the analysis, a list of identified hazards, their risk assessments, recommended actions, and a plan for follow-up.
9) Follow-up actions:
   - Review and prioritize the recommended actions based on the risk assessments.
   - Develop an action plan to implement the necessary changes, controls, or safeguards to mitigate identified hazards.
   - Assign responsibility for the action items and establish timelines for completion.
10) Review and update:
   - Periodically review and update the HAZOP study, especially when there are changes to the system, process, or operational conditions.
   - Ensure that the recommended actions are implemented and continue to be effective.

10.6.1.6 Results of a HAZOP Study

The results of a HAZOP are documented in a comprehensive report that summarizes the findings and recommendations of the study. The results of a HAZOP study are essential for enhancing process safety and risk management. The HAZOP report is a valuable tool for decision-makers, safety professionals, and regulatory authorities in ensuring the safe and reliable operation of industrial processes and systems. They provide a structured and documented analysis of potential hazards, enabling organizations to take proactive measures to mitigate risks and prevent accidents. Here are the key components and results typically included in a HAZOP study report:

(i) Scope and objectives: The report begins with a clear statement of the study's scope and objectives. It defines what was analyzed, the reasons for conducting the HAZOP, and the intended outcomes.
(ii) Team composition: It lists the names and roles of the participants who took part in the HAZOP sessions, including their areas of expertise and affiliations.
(iii) Node description: A detailed description of the system, process, or operation under study is provided. This includes information such as process flow diagrams, P&IDs, equipment specifications, and any relevant documentation to help readers understand the context.
(iv) Guidewords and parameters: The guidewords selected for the analysis are listed, along with the process parameters to which they were applied.
(v) Deviations and causes: The report presents a comprehensive list of identified deviations from the intended design or operational parameters. For each deviation, the report describes its causes, potential consequences, and severity.
(vi) Risk assessment: A risk assessment is performed for each identified hazard. This assessment typically considers factors like the severity of consequences, the likelihood of occurrence, and the detectability of the hazard. Risks are often categorized into low, medium, and high.
(vii) Safeguards and recommendations: The HAZOP report outlines the existing safeguards and controls in place to mitigate or prevent hazards. It also includes recommendations for additional safeguards or improvements to existing ones, based on the risk assessments. Recommendations are typically prioritized by risk level.
(viii) Action plan: An action plan is developed to address the recommended actions. This plan includes details such as who is responsible for each action, deadlines for implementation, and any required resources.
(ix) Summary of findings: The report provides a concise summary of the main findings, highlighting the most critical

hazards and their associated risks. This section is often useful for quick reference.

(x) HAZOP report approval: The report includes a section for signatures or approvals from key stakeholders, indicating their acknowledgment and acceptance of the study's findings and recommendations.

(xi) Appendices: Any additional supporting documentation, such as detailed calculations, drawings, or data, may be included in the appendices.

(xii) Review and update schedule: The report may specify a schedule for reviewing and updating the HAZOP study in the future, especially if the system or process is subject to changes.

10.6.1.7 The Role of Labeling in HAZOP

Labeling plays a crucial role in HAZOP by providing clear and standardized information about equipment, systems, and processes. Clear and consistent labeling of equipment, systems, and parameters ensures that the HAZOP team can accurately identify and assess potential hazards and deviations, ultimately contributing to safer and more efficient industrial processes. While labeling itself is not a formal part of the HAZOP process, it is an integral support mechanism that enhances the effectiveness and accuracy of HAZOP studies in several ways:

(i) Identification of equipment and components: Labels on equipment and components help HAZOP team members identify specific items in the process. Clear and consistent labeling ensures that everyone involved in the study understands what each element represents.

(ii) Process flow diagrams (PFDs) and piping and instrumentation diagrams (P&IDs): PFDs and P&IDs are essential tools in HAZOP studies. Equipment and lines in these diagrams are typically labeled with symbols and alphanumeric codes. Accurate labeling ensures that team members can refer to these diagrams easily during the analysis.

(iii) Node definition: In the preparation phase of a HAZOP study, the identification and definition of the "node" (the system, process, or operation under study) often rely on labeling. Properly labeled boundaries and components help specify the scope of the analysis.

(iv) Parameter identification: during the HAZOP analysis, team members examine various process parameters (e.g., temperature, pressure, and flow rate) for potential deviations. Labels on instruments and equipment provide essential context for identifying and discussing these parameters.

(v) Documentation: Accurate labeling supports the documentation of deviations, causes, consequences, and recommended safeguards during the HAZOP sessions. It ensures that the recorded information is associated with the correct equipment and process components.

(vi) Risk assessment: When assessing risks and consequences, team members rely on labels to associate identified hazards with specific components. This ensures that risk assessments are precise and properly attributed.

(vii) Follow-up actions: After the HAZOP study, recommended actions often include changes to equipment, instruments, or processes. Clear labeling aids in implementing these changes effectively and ensuring that the correct modifications are made.

(viii) Communication: Labels help facilitate communication and understanding among team members with different backgrounds and expertise. They provide a common language for discussing equipment and processes.

(ix) Safety and compliance: Proper labeling of equipment and systems is essential for safety and regulatory compliance. It ensures that personnel can easily locate and understand safety-related information, such as warning labels or emergency shutdown procedures.

(x) Training and awareness: Labels contribute to training and awareness efforts by providing visual cues and references for operators and maintenance personnel. Effective labeling can help prevent accidents and incidents.

10.6.2 Fault Tree Analysis (FTA)

A fault tree analysis (FTA) is a diagram-based, systematic technique for problem solving, troubleshooting, and determining the underlying cause of a failure. FTA is a top-down technique to issue solutions in which the undesirable occurrence serves as the starting point. The events that led up to the undesirable occurrence are subsequently assessed according to their relationship with their immediate consequence, indicating either another occurrence closer to the undesirable event or the problem itself. FTA may be used to investigate a single failure or to systematically investigate a collection of components, making it an adaptable instrument for root cause investigation. The primary use of FTA is in reliability and security engineering. Although FTA is crucial and well-known in the aerospace sector, it may be used by any company that wants to examine the vulnerability of its infrastructure to failure-causing events [7].

10.6.2.1 Purpose of a Fault Tree Analysis

The primary purpose of FTA is to provide a systematic and structured approach to understanding, quantifying, and managing the risks associated with undesired events and failures in complex systems. It supports safety, reliability, compliance, and decision-making processes in various industries. FTAs are used in various industries to achieve several key objectives:

(i) Risk assessment: FTA is a powerful tool for evaluating the risks associated with a specific undesired event or failure. It quantifies the probability of the event occurring based on the probabilities of its contributing causes. This enables organizations to prioritize and allocate resources for risk mitigation.

(ii) Safety improvement: By identifying the root causes and factors leading to undesired events, FTA helps organizations implement safety measures, improvements, and safeguards to prevent or mitigate these events. It contributes to enhancing safety in industrial processes and systems.

(iii) Reliability enhancement: FTA is used to assess the reliability of systems and identify vulnerabilities. Organizations can use the results to make design changes, implement redundancy measures, or improve maintenance practices to enhance the reliability of critical systems.

(iv) Accident investigation: In the aftermath of accidents or failures, FTA is employed as an investigative tool to determine the sequence of events and the root causes that led to the incident. It helps organizations understand what went wrong and why.

(v) Regulatory compliance: Many industries are subject to regulatory requirements related to safety and reliability. FTA can help organizations demonstrate compliance with these regulations by providing a systematic and quantitative assessment of risks and safety measures.

(vi) Design and process improvement: FTA can be used during the design or modification phase of systems and processes to proactively identify potential failure modes and vulnerabilities. This allows for the implementation of design changes or operational procedures that reduce the risk of undesired events.

(vii) Decision support: FTA provides decision-makers with valuable insights into the critical factors and causes contributing to an undesired event. This information supports informed decision-making regarding resource allocation, risk acceptance, and risk reduction strategies.

(viii) Emergency planning: FTA can be used to assess the potential impact of specific undesired events and to develop emergency response plans and procedures. It helps organizations prepare for and mitigate the consequences of incidents.

(ix) Quality control: FTA is sometimes used in quality control processes to identify potential sources of defects or failures in products or manufacturing processes. It can guide improvements in product design and manufacturing practices.

(x) Documentation and communication: FTA results are often documented in a visual and structured format, making it easier to communicate complex risk assessments and safety issues to stakeholders, including management, regulatory authorities, and personnel.

10.6.2.2 What are the Advantages of Performing a Fault Tree Analysis?

Performing a FTA offers several advantages and disadvantages, which should be carefully considered when deciding whether to use this method for risk assessment and analysis. Organizations should carefully consider the trade-offs between the advantages and disadvantages of FTA when deciding whether to use it for a particular risk assessment. Additionally, FTA is often most effective when used in conjunction with other risk assessment methods and qualitative insights. Here are the key advantages and disadvantages of conducting a FTA:

Advantages of FTA:

- Systematic analysis: FTA takes an organized and scientific method when analyzing complex systems and events, making sure that all possible contributing aspects are taken into account.
- Identification of root causes: FTA helps identify the root causes of undesired events or failures, allowing organizations to address the underlying issues rather than just treating symptoms.
- Quantitative assessment: FTA allows for quantitative assessment of risk by assigning probabilities to events and their combinations. This enables prioritization of risks and allocation of resources for risk mitigation.
- Visual representation: The fault tree diagram provides a clear and visual representation of the relationships between events and factors contributing to an undesired event, making it easier to understand and communicate the analysis results.
- Early detection of risks: FTA can be used during the design phase of a system or process to proactively identify and address potential failure modes and vulnerabilities.
- Risk mitigation: FTA results help organizations implement safety measures, improvements, and safeguards to prevent or mitigate the identified risks, contributing to enhanced safety and reliability.
- Objective decision support: FTA gives objective statistics to decision-makers so that they may make educated decisions about risk acceptance, risk-reduction methods, and resource allocation.
- Regulatory compliance: FTA can be used to demonstrate compliance with safety and reliability regulations and standards in various industries.

Disadvantages of FTA:

- Complexity: FTA may become quite complicated, particularly for systems with numerous components and various possible failure mechanisms. Controlling such complexity can be difficult.

- Resource-intensive: Performing a detailed FTA requires a significant investment of time, expertise, and resources. Data collection and analysis can be time-consuming.
- Data availability: Accurate and reliable data on event probabilities and causes may not always be readily available, leading to uncertainty in the analysis.
- Subjectivity: The assignment of probabilities and assessments of events may involve subjective judgments, introducing a degree of uncertainty into the analysis.
- Limited scope: FTA may focus primarily on technical or engineering aspects and may not fully consider human factors, organizational issues, or external events that can contribute to undesired events.
- Combinatorial explosion: As the number of events and components increases, the number of possible event combinations in a fault tree can grow exponentially, making analysis more complex.
- Static analysis: FTA is typically a static analysis method, meaning it does not account for dynamic changes and interactions that can occur in real-time systems.
- Overemphasis on quantification: Excessive emphasis on quantitative analysis can lead to an overreliance on numerical values, potentially overlooking critical qualitative insights.

10.6.2.3 Steps for Creating Fault Tree Analysis Diagram

Creating a FTA diagram involves several steps to systematically represent the relationships between events and factors contributing to an undesired event or failure. Creating a FTA diagram requires careful planning, domain knowledge, and attention to detail. It is a valuable tool for understanding and managing complex risks in various industries. Here is a step-by-step guide on how to create a FTA diagram:

Step 1: Define the top event: Start by clearly defining the top event, which is the undesired event or failure that should be analyzed. This event is placed at the top of the fault tree diagram.
Step 2: Identify basic events: Identify and list the basic events that can contribute to the top event. These should be specific events or conditions that can occur and are related to the top event.

Step 3: Determine logic gates: Choose the appropriate logic gates (AND and OR gates) to represent the relationships between events. Logic gates determine how the basic events are combined to contribute to the top event. Use an AND gate when all input events must occur for the output event to occur. This represents a logical "AND" relationship and is typically represented as an inverted "V." Use an OR gate when at least one input event must occur for the output event to occur. This represents a logical "OR" relationship and is typically represented as an inverted "U." Table 10.1 depicts the different symbols used for drawing the FTA diagram.

Table 10.1 Different symbols used in FTA.

Name	Symbol	Description
Basic Event		The primary or fundamental failure event. It is a random event, and enough data is accessible.
Intermediate Event		System, subsystem, or component event state.
Undeveloped Event		Secondary failure or an undeveloped incident might be investigated further.
Conditioning Event		A conditional event occurs as a result of the occurrence of another event.
External Event		The occurrence or non-occurrence of an event is represented by a house event.
Transfer Event	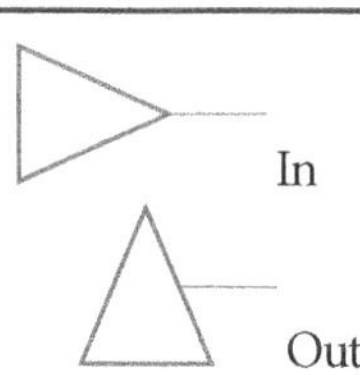	Transfer-in and transfer-out marks are used to recreate an FTA branch or subtree.
AND Gate		When all of the input events occur, the output event happens.
OR Gate	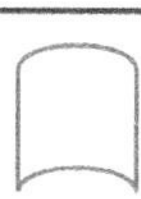	Whenever less than one of the input's events happens, the output event occurs.

(Continued)

Table 10.1 Different symbols used in FTA. (*Continued*)

Name	Symbol	Description
Priority AND Gate		When all of those input occurrences occur in the correct sequence from left to right, the output event happens.
Exclusive OR Gate	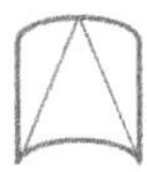	If any of those two input events happen but not both, then the output event occurs.
Inhibit Gate		When an input event happens and the linked condition is met, the output event occurs.

Step 4: Create the fault tree diagram: Start by drawing the top event at the top of the diagram. Connect the basic events to the top event using the chosen logic gates. Indicate whether each gate is an AND or OR gate. Continue branching out and connecting events, intermediate events (if used), and logic gates until all relevant causes and contributing factors are represented.

Step 5: Label the diagram: Label each event, including the top event, basic events, and intermediate events, with clear and descriptive names. Add numerical values or probabilities to the basic events, if available, to represent the likelihood of each event occurring.

Step 6: Review and verify: Review the fault tree diagram to ensure that it accurately represents the relationships between events and that it is logically structured.

Step 7: Verify that the logic gates and event combinations align with the intended analysis.

Step 8: Documentation: Document the FTA, including the fault tree diagram, probabilities (if used), and any relevant notes or observations.

Step 9: Interpret results: Interpret the results of the FTA to understand the critical factors and causes contributing to the top event. This information is valuable for risk assessment and risk management.

Step 10: Implement risk mitigation: Based on the analysis, develop strategies to mitigate the identified risks, enhance safety measures, or implement improvements to prevent or reduce the occurrence of the top event.

10.6.3 Event Tree Analysis (ETA)

Event tree analysis (ETA) is a methodical and graphical approach for analyzing the potential outcomes or repercussions of a starting point event, generally in the context of risk evaluation and safety analysis. To examine the sequence of events that might occur from an initial incident or disaster, ETA is extensively used in sectors such as nuclear power, aviation, chemical processing, and oil and gas. Event tree analysis is an organized and methodical technique to evaluate the implications of starting events that assists organizations in making educated decisions to improve safety, risk reduction, and incident prevention [8]. Key concepts in ETA:

- Initiating event: The initiating event is the starting point of the analysis. It represents an event or condition that triggers a chain of events or decisions.
- Intermediate events: Intermediate events are events or conditions that occur in response to the initiating event. They can be either "success" events (leading to a positive outcome) or "failure" events (leading to a negative outcome).
- End states: End states are the final outcomes or consequences of the event sequence. These can include various levels of severity, such as minor incidents, accidents, or catastrophic events.
- Branches: Branches represent possible paths or sequences of events that follow the initiating event. They illustrate how different events and decisions lead to various end states.

10.6.3.1 The Steps in Event Tree Analysis

Event tree analysis is often used in risk assessment and safety analysis in various industries for evaluating the possible consequences of events and helps organizations make informed decisions regarding risk management, safety enhancements, and incident prevention. Here are the steps involved in ETA:

(i) Define the initiating event: Clearly define the initiating event or scenario that one want to analyze. This is the event that starts the sequence.

(ii) Identify intermediate events: Identify and list the possible intermediate events or conditions that can occur as a result of the initiating event. These events can be categorized as "success" or "failure."

(iii) Develop the Event Tree Diagram: Create a graphical representation of the event sequence by arranging the initiating event at the top and branching out to show the different intermediate events and their relationships. Use arrows or lines to connect events and show the sequence.

(iv) Assign probabilities: Assign probabilities to each intermediate event, reflecting the likelihood of it occurring as a result of the preceding event or decision. Probabilities can be based on historical data, expert judgment, or calculations.

(v) Evaluate end states: For each path in the event tree, consider the potential end states or outcomes. These can range from minor incidents to major accidents, depending on the sequence of events.

(vi) Quantitative analysis: If desired, perform quantitative analysis to calculate the overall probability of reaching specific end states. This can involve combining probabilities along different branches.

(vii) Risk assessment: Assess the risks associated with each end state, considering the consequences, likelihood, and potential mitigation measures.

(viii) Documentation: Document the entire ETA process, including the event tree diagram, assigned probabilities, and analysis results.

(ix) Interpret results: Interpret the results of the analysis to understand the likelihood and consequences of different end states. This information helps in making informed decisions and managing risks.

(x) Mitigate risks: Develop strategies to mitigate risks and prevent or minimize the occurrence of undesirable end states. These strategies may include safety measures, preventive actions, or emergency response plans.

10.6.3.2 Advantages and Disadvantages of Event Tree Analysis

ETA is a valuable tool for assessing the consequences of initiating events and making informed decisions about risk management. However, it is essential to be aware of its limitations, particularly regarding data availability, complexity, and subjectivity, and to consider these factors when applying ETA to specific scenarios.

Advantages of ETA:

- Structured analysis: ETA provides a structured and systematic approach to assessing the sequence of events and their consequences, ensuring that all possible outcomes are considered.
- Visual representation: The use of graphical diagrams makes complex event sequences easy to understand and communicate to stakeholders, including non-technical audiences.
- Scenario exploration: ETA allows for the exploration of various scenarios and end states resulting from initiating events, helping organizations plan for a wide range of possibilities.
- Probability assessment: ETA enables the assignment of probabilities to intermediate events, allowing for quantitative analysis of the likelihood of reaching specific end states.
- Risk assessment: ETA helps organizations assess the risks associated with different end states, taking into account consequences and probabilities, which is essential for risk management.
- Decision support: The analysis results can inform decision-makers about the potential outcomes of different choices and support informed decision-making.
- Preventive measures: By identifying potential end states and their likelihood, organizations can implement preventive measures and safety measures to reduce the risk of undesirable outcomes.
- Safety improvement: ETA can lead to safety improvements and enhancements in operational procedures, leading to safer practices and processes.

Disadvantages of ETA:

- Complexity: ETA can become highly complex, especially for scenarios with multiple branching sequences. Managing and interpreting complex event trees can be challenging.
- Data availability: Reliable data, especially probabilities for intermediate events, may not always be readily available. This can introduce uncertainty into the analysis.
- Subjectivity: Assigning probabilities and assessing the likelihood of events may involve expert judgment, leading to subjectivity in the analysis.
- Deterministic assumption: ETA often relies on deterministic modeling, assuming that events and decisions have fixed probabilities and outcomes, which may not always reflect real-world variability.
- Limited focus: ETA primarily focuses on the sequence of events and their consequences, often overlooking other factors such as human error, organizational culture, and external factors that can influence outcomes.
- Resource-intensive: Developing and conducting a detailed ETA analysis can be resource-intensive, requiring time, expertise, and access to data.
- Quantitative challenges: Quantitative analysis in ETA may be challenging, particularly when dealing with complex or rare events where data is limited.
- Dynamic factors: ETA is typically a static analysis method and may not capture dynamic changes and interactions that occur in real-time systems.

10.6.4 Preliminary Hazard Analysis (PHA)

Preliminary hazard analysis is the initial phase in the systematic process of identifying and evaluating possible risks in a system, process, or project. It is often performed early in the development of a project or system to offer a wide overview of possible hazards. Prior to more extensive assessments, PHA seeks to identify and prioritize dangers [9]. It is a semi-quantitative study that is used to identify all potential hazards and unintentional events that might lead to an accident, rate the detected accidental events based on their severity, and determine essential hazard controls and follow-up activities.

10.6.4.1 *Key Components of Preliminary Hazard Analysis*

(i) System or process description: Begin by describing the system, process, or project under consideration. This includes defining its purpose, components, interfaces, and any relevant context.

(ii) Identify hazards: Identify potential hazards associated with the system, process, or project. Hazards can include physical hazards, chemical hazards, environmental hazards, operational hazards, and more.

(iii) Hazard causes: Determine the root causes or sources of each identified hazard. This involves considering events, conditions, or actions that could lead to the occurrence of the hazard.

(iv) Consequences: Assess the potential consequences of each hazard. Consider the impact on personnel, the environment, equipment, and the overall operation if the hazard were to occur.

(v) Risk assessment: Perform a preliminary risk assessment by evaluating the likelihood and severity of each hazard. This can be done qualitatively, semi-quantitatively, or quantitatively, depending on the available data and the level of detail required.

(vi) Risk prioritization: Prioritize the identified hazards based on their level of risk. This helps determine which hazards require further analysis and mitigation efforts.

(vii) Mitigation strategies: Propose preliminary mitigation strategies or controls for the most significant hazards. These strategies may involve design changes, operational procedures, safety measures, or other actions to reduce or eliminate the risk.

(viii) Documentation: Document the results of the PHA, including a list of identified hazards, their causes, consequences, risk assessments, and proposed mitigation measures.

10.6.4.2 *Steps to Carrying Out Preliminary Hazard Analysis*

Carrying out a preliminary hazard analysis is a proactive step in identifying and addressing potential risks in a systematic and structured manner. Preliminary hazard analysis involves a structured process to identify and assess potential hazards in a system, process, or project. It lays the foundation for more detailed hazard analysis and safety improvements. Here are the steps to perform a PHA:

(i) Define the scope: Clearly identify the scope of the study. Determine the specific structure, process, or project to be examined. Determine the analysis' boundaries and interfaces.

(ii) Assemble a cross-functional team: Form a team of experts and stakeholders with diverse backgrounds and expertise relevant to the system or process under analysis. This may include engineers, safety professionals, operators, and others.

(iii) System or process description: Provide a detailed description of the system, process, or project. Include information about its purpose, components, interfaces, and any relevant context. A clear understanding of the system is essential for hazard identification.

(iv) Hazard identification: Identify potential hazards associated with the system or process. Hazards can encompass various categories, such as physical, chemical, operational, and environmental. Use brainstorming sessions, historical data, and available documentation to identify hazards.

(v) Hazard causes: Determine the root causes or sources of each identified hazard. Explore the events, conditions, or actions that could lead to the occurrence of these hazards. Understanding the causes is crucial for developing effective mitigation strategies.

(vi) Consequences analysis: Assess the potential consequences of each identified hazard. Consider the impact on personnel, the environment, equipment, and the overall operation if the hazard were to occur. Categorize consequences by severity.

(vii) Risk assessment: For each danger, conduct a preliminary risk assessment. This entails assessing the chance of the danger occurring as well as the severity of its repercussions. Depending on the available data and resources, risk assessment can be done qualitatively or semi-quantitatively.

(viii) Risk prioritization: Prioritize the identified hazards based on their risk levels. Typically, prioritize hazards with higher risks for further analysis and mitigation efforts.

(ix) Mitigation strategies: Propose preliminary mitigation strategies or controls for the most significant hazards. These methods might include modifications to the design, operating processes, safety measures, or other activities to decrease or eliminate the risk associated with each hazard. Ensure that the proposed measures are feasible and effective.

(x) Documentation: Document the results of the PHA in a clear and organized manner. Create a comprehensive record that includes a list of identified hazards, their causes, consequences, risk assessments, and proposed mitigation measures. Well-organized documentation is crucial for reference and decision-making.

(xi) Review and validation: Consider conducting a review and validation process to ensure the accuracy and completeness of the PHA results. Input from subject matter experts and stakeholders can help refine the analysis and identify any overlooked hazards or mitigation opportunities.

(xii) Periodic reassessment: Recognize that the hazard landscape may change over time. Plan for periodic reassessments of the PHA to account for evolving risks, changes in the system or process, and the effectiveness of implemented mitigation measures.

(xiii) Reporting and communication: Share the results of the PHA with relevant stakeholders, including decision-makers, engineers, operators, and safety professionals. Effective communication of the analysis findings is crucial for implementing mitigation measures and improving safety.

10.6.4.3 Advantages of Preliminary Hazard Analysis

- Early identification: PHA identifies potential hazards at an early stage, allowing organizations to address safety concerns before they become more challenging and costly to mitigate.
- Risk awareness: It raises awareness about potential risks among project stakeholders and decision-makers, fostering a safety-conscious culture.
- Cost-effective: PHA can help prevent costly accidents, damage, or delays by addressing hazards early in the project lifecycle.

- Risk prioritization: It helps prioritize risks for further analysis and targeted mitigation efforts, ensuring that resources are allocated effectively.

10.6.4.4 *Disadvantages and Limitations of PHA*

- Limited detail: PHA provides a high-level overview and may not identify all potential hazards or consider complex interactions among factors.
- Subjectivity: Risk assessments in PHA may be subjective, depending on the expertise and judgment of the analysis team.
- Incomplete data: In the early stages of a project, there may be limited data available to assess risks accurately.
- Evolution of hazards: Hazards may evolve or change as the project progresses, making it necessary to revisit the analysis periodically.
- Resource-intensive: Conducting PHA requires time, expertise, and resources, which may not always be available in some projects.

10.6.5 Bow Tie Analysis

Bow tie analysis is a risk assessment and management technique used to visualize and analyze potential hazards, their causes, consequences, and preventive and mitigative controls in complex systems or processes. It is named for its visual representation, which resembles a bow tie. Bow tie analysis creates a visual representation of a hazard by using a "bow tie" diagram. It identifies the hazard in the center, the causes on the left, and the consequences on the right. It helps in visualizing and communicating the risk controls. Bow tie analysis is widely used in industries such as aviation, oil and gas, healthcare, and chemical processing to assess and manage risks comprehensively [10]. Figure 10.2 depicts the process of bow tie analysis.

10.6.5.1 *Key Components of Bow Tie Analysis*

(i) Top event: The top event represents the undesired incident or hazard one want to assess. It is placed at the center of the bow tie diagram.

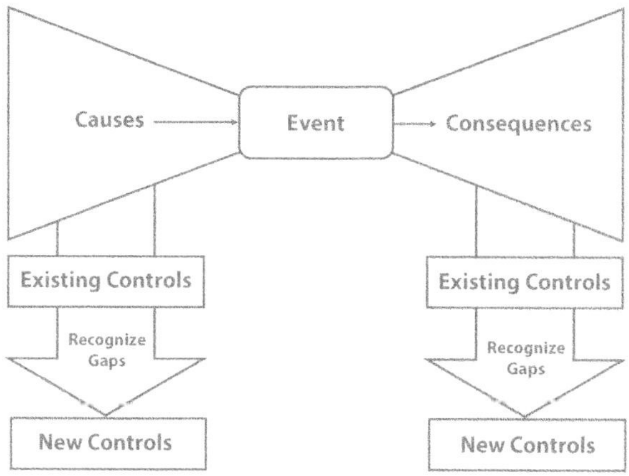

Figure 10.2 Bow tie analysis.

(ii) Threats or causes: On the left side of the diagram, one identify threats or causes that could lead to the top event. These threats are the potential triggers of the hazard.
(iii) Consequences: On the right side of the diagram, one identify consequences that may result from the top event. These consequences are the potential impacts or outcomes of the hazard.
(iv) Barriers or controls: Above and below the top event, one identify barriers or controls. These are preventive and mitigative measures that are put in place to either reduce the likelihood of the top event occurring (preventative controls) or minimize the impact if it does occur (mitigative controls).

10.6.5.2 The Steps in Bow Tie Analysis

Bow tie analysis is a risk assessment and management technique that visualizes and analyzes potential hazards and the controls in place to mitigate those hazards. The key components of Bow tie analysis include the following:

(i) Define the hazard: Clearly define the hazard or top event one want to analyze. It should be specific and well-understood.

(ii) Identify threats and consequences: Identify and list the threats or causes that could lead to the top event on the left side of the diagram. On the right side, list the potential consequences of the top event.

(iii) Identify barriers and controls: Above and below the top event, list the barriers or controls that are in place or should be implemented to prevent or mitigate the hazard. These controls can be administrative, engineering, or procedural in nature.

(iv) Assess the likelihood and severity: Qualitatively or quantitatively assess the likelihood of each threat causing the top event and the severity of the consequences if it occurs.

(v) Evaluate barriers and controls: Evaluate the effectiveness of the existing barriers and controls in preventing or mitigating the hazard. Identify any gaps or weaknesses in the control measures.

(vi) Risk assessment: Combine the likelihood and severity assessments to determine the overall risk associated with the hazard. This helps prioritize the hazard for further action.

(vii) Risk mitigation: Develop and implement strategies to strengthen existing controls, add new controls, or revise procedures to reduce the risk to an acceptable level.

(viii) Documentation: Document the entire bow tie analysis, including the diagram, assessments, identified controls, and risk mitigation measures.

10.6.5.3 Advantages and Disadvantages of Bow Tie Analysis

Bow tie analysis is a valuable tool for understanding, visualizing, and managing risks in complex systems that not only offers several advantages but also comes with certain disadvantages and limitations. Its advantages lie in its ability to provide a holistic view of hazards and controls, facilitate communication, and support proactive risk management. However, it also has limitations related to complexity, data requirements, subjectivity, and the need for ongoing maintenance and updates. Organizations should consider these factors when deciding whether to use bow tie analysis for risk assessment and management.

Advantages of Bow Tie Analysis:

- Visual clarity: The bow tie diagram provides a clear and intuitive visual representation of the hazard, its causes, consequences, and control measures. This visual clarity makes it easier for stakeholders to understand and communicate risks.
- Comprehensive risk assessment: Bow tie analysis allows for a comprehensive assessment of both preventive and mitigative controls. It considers not only the causes and consequences of a hazard but also the measures in place to manage it.
- Effective communication: The visual nature of the bow tie diagram makes it an effective tool for communicating risk information to various stakeholders, including management, employees, and regulatory authorities.
- Root cause analysis: Bow tie analysis can be used to explore the root causes of hazards and incidents. It helps in understanding the underlying factors that contribute to the top event.
- Proactive risk management: By identifying threats, consequences, and control measures, bow tie analysis promotes proactive risk management. It enables organizations to implement preventive and mitigative actions to reduce risk.
- Scenario exploration: Bow tie diagrams can be adapted to analyze different scenarios and variations, allowing organizations to explore a wide range of potential outcomes and risks.
- Auditing and compliance: Bow tie diagrams serve as valuable tools for auditing safety and compliance with regulatory requirements. They provide a visual record of risk assessment and management efforts.

Disadvantages and Limitations of Bow Tie Analysis:

- Complexity: For complex systems or scenarios, bow tie diagrams can become intricate and challenging to manage. The complexity increases as more threats, consequences, and controls are considered.
- Data requirements: Gathering data for quantitative risk assessments in bow tie analysis can be resource-intensive

and may require access to accurate and comprehensive data, which may not always be readily available.

- Subjectivity: Some aspects of the analysis, such as likelihood and consequence assessments, may involve subjective judgment, potentially leading to variability in results.
- Resource-intensive: Creating and maintaining bow tie diagrams can be resource-intensive in terms of time, expertise, and documentation efforts.
- Limited predictive capability: While bow tie analysis is effective for analyzing existing risks and controls, it may have limited predictive capability for novel or emerging risks that have not been previously identified.
- Inadequate consideration of human factors: Bow tie analysis tends to focus on technical and system-related aspects, often overlooking the role of human factors, including human error and behavior, which can be significant contributors to risk.
- Static analysis: Bow tie analysis is typically a static analysis method and may not capture dynamic changes and interactions that occur in real-time systems.

10.6.6 Failure Modes and Effects Analysis (FMEA)

Failure modes and effects analysis is a tool for identifying, prioritizing, and limiting failure modes in a process. It aids in continuous development by documenting current failure risk information and actions. FMEA is used during design to prevent failures and thereafter for control prior to and during process operation. It is best employed in the early conceptual stages of design and continues throughout the life of the product or service. FMEA does not replace excellent engineering but rather improves it by utilizing the knowledge and expertise of a cross-functional team to analyze the design and development of a product or process and estimate its risk of failure [11].

10.6.6.1 Different Types of FMEA

FMEA can be categorized into several different types or variations, each tailored to specific applications and industries. Each type of FMEA is tailored to the specific context and objectives of a project or industry. The choice of which type to use depends on the stage of development, the nature of the system or product, and the primary areas of concern for preventing failures and improving quality. The main types of FMEA include the following:

(i) Design FMEA (DFMEA):
- Focuses on identifying and mitigating potential failure modes and their effects during the design and development phase of a product or system.
- Aims to prevent design-related issues from reaching the manufacturing or production stage.
- Evaluates factors such as materials, specifications, tolerances, and design choices.

(ii) Process FMEA (PFMEA):
- Concentrates on analyzing potential failure modes and their effects within a manufacturing or production process.
- Aims to identify and address issues related to the manufacturing process, equipment, human factors, and environmental factors.
- Helps improve product quality and reduce defects during production.

(iii) System FMEA (SFMEA):
- Examines the potential failure modes and their effects at the system level, considering interactions between different subsystems or components.
- Useful for complex systems or projects where multiple elements must work together seamlessly.

(iv) Software FMEA (SW-FMEA):
- Specifically designed for software development and focuses on identifying and addressing failure modes related to software systems.
- Aims to prevent software bugs, glitches, and vulnerabilities that can affect system performance or security.

(v) Functional FMEA (FFMEA):
- Analyzes the potential failure modes and their effects with a primary focus on the intended functions of a product or system.
- Helps ensure that a product or system meets its functional requirements and objectives.

(vi) Hardware FMEA (HW-FMEA):
- Similar to DFMEA but with a narrower focus on the hardware components of a product or system.
- Evaluates the potential failure modes and their effects on hardware elements such as electronic components, mechanical parts, or structural elements.

(vii) Maintenance FMEA (MFMEA):
- Addresses potential failure modes and their effects during the maintenance or servicing phase of a product or system's life cycle.
- Aims to ensure that maintenance activities are effective, safe, and cost-efficient.

(viii) Use FMEA (UFMEA):
- Focuses on identifying and mitigating potential failure modes and their effects during the operational or user phase.
- Aims to improve user safety, product reliability, and customer satisfaction.

(ix) Supplier FMEA (SFMEA):
- Analyzes potential failure modes and their effects related to components or materials supplied by external vendors or suppliers.
- Helps manage supply chain risks and improve the quality of purchased components.

10.6.6.2 Why Perform Failure Mode and Effects Analysis (FMEA)

Performing FMEA offers numerous benefits in various industries and contexts. It plays a crucial role in various industries by helping organizations identify and address potential failures and their consequences, ultimately leading to more reliable, safe, and cost-effective products and processes. Here are some key reasons why organizations and professionals use FMEA:

- Risk identification and mitigation: FMEA systematically identifies potential failure modes, their causes, and their effects. By doing so, it helps organizations proactively recognize and mitigate risks associated with products, processes, or systems.
- Quality improvement: FMEA is a valuable tool for enhancing product and process quality. By addressing potential failure modes and their causes early in the design or

production phases, organizations can reduce defects, errors, and non-conformities.

- Enhanced safety: FMEA helps in identifying failure modes that could lead to safety hazards or accidents. By addressing these issues, organizations can enhance the safety of their products and operations, reducing the risk of harm to users and employees.
- Cost reduction: Preventing failures through FMEA can lead to cost savings. By addressing potential issues early in the development or production process, organizations can avoid expensive rework, recalls, and warranty claims.
- Compliance and regulation: FMEA is often a requirement in industries with strict regulatory standards, such as aerospace, automotive, and healthcare. Performing FMEA helps organizations meet compliance requirements and demonstrate their commitment to quality and safety.
- Customer satisfaction: Delivering reliable and high-quality products or services is essential for maintaining customer satisfaction. FMEA helps identify and address issues that could negatively impact customer experience and reputation.
- Decision support: FMEA provides valuable information for decision-making. It helps organizations prioritize which failure modes to address first based on factors like severity, occurrence, and detection, enabling more informed decisions about risk mitigation.
- Continuous improvement: FMEA is not something that happens once. It fosters a culture of constant enhancement by encouraging organizations to examine and update their analyses on a frequent basis when fresh knowledge becomes available or as things or procedures change.
- Efficiency and productivity: Identifying and mitigating potential failure modes can lead to more efficient and productive operations. It reduces downtime, rework, and the need for emergency fixes.
- Supplier and vendor management: FMEA can be used to assess the risks associated with components or materials supplied by external vendors. This helps organizations manage supply chain risks and select reliable suppliers.
- Research and development: In research and development contexts, FMEA can be used to evaluate new technologies or

materials for their potential failure modes and effects, allowing for informed decisions in the development process.

- Project management: FMEA can be applied to project management to identify risks and challenges that could impact project success. This helps project managers plan for contingencies and allocate resources effectively.

10.6.6.3 When to Perform FMEA

FMEA is a valuable tool that can be performed at various stages of a product's lifecycle or during different phases of a project. The timing of when to perform FMEA depends on the specific context and objectives of the analysis. The specific timing of FMEA activities may vary by industry, organization, and project. However, the general principle is to start FMEA as early as possible in the development or project lifecycle to proactively identify, assess, and mitigate potential failures, thereby reducing risks, improving quality, and enhancing safety. Here are common instances when FMEA should be conducted:

1. Early design phase (DFMEA):
   - New product development: Perform DFMEA during the initial design stages of a new product to identify and mitigate potential design-related failures before manufacturing begins.
   - Redesign or product updates: Conduct DFMEA when making significant changes or updates to an existing product design to assess the impact on reliability and safety.
2. Process planning phase (PFMEA):
   - Process development: Perform PFMEA when developing or modifying manufacturing or production processes to identify and address potential failure modes related to the process itself.
   - Production startup: Conduct PFMEA before launching a new production line or process to ensure smooth operations and minimize defects.
3. System integration phase (SFMEA):
   - Integration of subsystems: Use SFMEA when integrating different subsystems or components into a larger system to assess the interactions and potential failure modes at the system level.

- Complex projects: In complex projects involving multiple systems or disciplines, SFMEA helps manage overall project risks.

4. Software development phase (SW-FMEA):
   - Software design: Perform SW-FMEA during the software design and development process to identify potential software-related failures and vulnerabilities.
   - Software updates: Conduct SW-FMEA when releasing software updates or patches to assess the impact on system performance and security.
5. Operations and maintenance phase (UFMEA and MFMEA):
   - Product deployment: Use UFMEA when deploying products or systems in real-world operations to anticipate potential issues and ensure safety and reliability.
   - Maintenance planning: Conduct MFMEA when planning maintenance activities to prevent unexpected failures during routine or corrective maintenance.
6. Supplier and vendor selection (SFMEA):
   - Supplier evaluation: Use SFMEA to assess the risks associated with components or materials supplied by external vendors before selecting and contracting with suppliers.
7. Continuous improvement (Ongoing):
   - FMEA is not a one-time activity; it should be part of an ongoing quality and risk management process. Update and review FMEAs on a regular basis when fresh data becomes available, modifications are made to goods or processes, or problems are detected.
8. Project management (PFMEA):
   - Use PFMEA to identify project-related risks and challenges that could impact project success, especially in complex or critical projects.

10.6.6.4 How to Perform FMEA

Performing FMEA involves a systematic step-by-step process to identify and assess potential failure modes, their causes, and their effects. FMEA is a continuous improvement tool that helps organizations proactively manage risks and enhance quality. It should be integrated into the product development or process management lifecycle to ensure ongoing reliability and safety. Here is a general guide on how to perform FMEA:

(i) Assemble a cross-functional team: Form a team that includes individuals with diverse expertise related to the product, process, or system under analysis. This may include designers, engineers, quality assurance professionals, and subject matter experts.

(ii) Select the scope and boundaries: Define the scope of the FMEA, including what is being analyzed (e.g., a product, process, or system), its intended use, and any specific boundaries or limitations.

(iii) Identify the functions and requirements: List the primary functions and requirements of the product, process, or system. This step helps ensure that the analysis aligns with the intended purpose.

(iv) Identify potential failure modes: Brainstorm and list all the potential ways in which the product, process, or system could fail. Consider both functional failures (e.g., it does not perform its intended function) and non-functional failures (e.g., safety, reliability, or quality issues).

(v) Identify failure causes: For each recognized failure mode, identify the various causes or variables that might contribute to that failure. Consider all relevant aspects, including design, components, human error, and the environment.

(vi) Determine the effects of failure: Assess the potential consequences or effects of each failure mode on the product, process, or system. Consider factors like safety, quality, performance, and regulatory compliance. Each effect should be assigned a severity rating, often on a numerical scale (e.g., 1 to 10), with higher numbers indicating more severe repercussions.

(vii) Evaluate the severity likelihood: severity ratings are based on the risk's criticality. It assesses the effect of the failure mode. Severity ratings range from 1 to 10, with 1 indicating very low severity and 10 indicating great severity, resulting in risk without indication or warning. In certain sectors, categories are used rather than numerical numbers.

(viii) Estimate the occurrence likelihood: Calculate the possibility or likelihood of each failure mode happening. This evaluation should take into account past facts, experience, and expert judgement. Assign an occurrence rating on a numerical scale, with higher numbers indicating a higher likelihood of occurrence.

(ix) Estimate the likelihood of detection: Determine the likelihood of the failure mode being recognized before it affects the end user or creates damage. This assessment considers quality control measures, testing, inspections, and monitoring. Assign a detection rating on a numerical scale, with higher numbers indicating a lower likelihood of detection.

(x) Determine the risk priority number (RPN): RPN is obtained from the product of severity, incidence, and detection ratings for each failure mode. Table 10.2 depicts the FMEA scales for different ratings. Higher RPN values imply increased risk, which helps prioritize failure modes to tackle first.

(xi) Prioritize and develop action plans: Sort the failure modes by their RPN values, with the highest RPNs at the top of the list. Focus on addressing the high-risk failure modes first.

(xii) Implement and monitor actions: Put the action plans into practice and monitor their effectiveness in reducing the identified risks. Continuously review and update the FMEA as needed.

(xiii) Document the FMEA: Keep comprehensive records of the FMEA process, comprising the failure modes detected, their causes and consequences, ratings, RPN values, and response plans.

10.6.6.5 Advantages and Disadvantages of FMEA

FMEA is a powerful tool for risk assessment, quality improvement, and proactive problem-solving. However, like any methodology, it has both advantages and disadvantages. Despite its limitations, FMEA remains a valuable tool for organizations looking to proactively manage risks, enhance quality, and improve safety. When used appropriately and integrated into a comprehensive quality management system, FMEA can provide significant benefits. Let us explore these in more detail:

Advantages of FMEA:

- Risk identification and mitigation: FMEA examines probable failure modes, their causes, and their consequences in a methodical manner. This proactive strategy enables organizations to manage risks before they turn into serious

Table 10.2 FMEA scales.

Severity Rating (1 for least severe to 10 for most severe)	**Occurrence Rating (1 for least frequent to 10 for most frequent)**	**Detection Rating (1 for always detected to 10 for never detected)**
Minor (Rank 1): It is unreasonable to suppose that this minor failure would have any discernible impact on item or system performance or future processes or assembly operations. The malfunction will most likely go undetected by the customer.	Remote (Rank 1): Failure is quite unlikely. There have never been any failures linked to this or nearly equivalent processes.	Very High (Ranks 1–2): The failure mode will almost certainly be detected by current controls. Similar procedures have reliable detection controls. The procedure disables further processing automatically.
Low (Ranks 2–3): Because of the nature of this failure, the consumer simply perceives little inconvenience. The customer will most likely notice a little degradation in the item's or system's performance, as well as a minor discomfort with a later process or assembly operation, i.e., minimal rework.	Very Low (Rank 2): Only single failures linked with this technique or almost equivalent processes	High (Ranks 3–4): Controls have a strong likelihood of recognizing failure mode; the process recognizes failure mode automatically.

(Continued)

Table 10.2 FMEA scales. (*Continued*)

Severity Rating (1 for least severe to 10 for most severe)	**Occurrence Rating (1 for least frequent to 10 for most frequent)**	**Detection Rating (1 for always detected to 10 for never detected)**
Moderate (Ranks 4–6): Failure results in some level of client displeasure, which may include inconvenience or aggravation. The customer will perceive a decrease in item or system performance. This can result in unexpected rework, repair, and/or equipment damage.	Low (Ranks 3–5): Isolated failures caused by comparable mechanisms	Moderate (Ranks 5–6): A failure mode may be detected via controls.
High (Ranks 7–8): Customer discontent is high owing to the nature of the failure, such as an inoperable item or system. Failure has nothing to do with safety or government control. This may cause a significant interruption to the following processing or assembly activities and/or necessitate extensive rework.	Moderate (Ranks 6–7): This procedure has frequent failures, although not in significant numbers.	Low (Ranks 7–8): The controls have a low probability of identifying the presence of a failure mode.

(*Continued*)

Table 10.2 FMEA scales. (*Continued*)

Severity Rating (1 for least severe to 10 for most severe)	**Occurrence Rating (1 for least frequent to 10 for most frequent)**	**Detection Rating (1 for always detected to 10 for never detected)**
Very High (Ranks 9–10): Failure has a negative impact on safety or includes noncompliance with government standards. May jeopardize the machine or the assembly operator (9 with warning, 10 without).	High (Ranks 8–9): This or comparable techniques have frequently failed.	Very low (Rank 9): Controls are unlikely to identify the presence of a failure mode.
	Very High (Rank 10): Failure is nearly unavoidable.	Absolutely No Detection (Rank 10): Controls will not be able to identify the presence of a failure. There are no known controls to identify the failure mode.

concerns, lowering the possibility of failures and the costs that come with them.

- Improved product and process quality: FMEA aids in the improvement of product and process quality by detecting and correcting any shortcomings or deficiencies early in the research and development stages.
- Enhanced safety: FMEA is valuable for identifying failure modes that could pose safety hazards. By addressing these issues, organizations can enhance safety for users and employees.
- Cost reduction: Preventing failures through FMEA can lead to significant cost savings. It reduces the need for rework, warranty claims, and costly emergency fixes.

- Compliance and regulation: FMEA helps organizations meet regulatory and compliance requirements by demonstrating a systematic approach to risk management and quality improvement.
- Customer satisfaction: Delivering reliable and high-quality products or services improves customer satisfaction and enhances an organization's reputation.
- Decision support: FMEA provides valuable information for decision-making by prioritizing which failure modes to address first based on risk factors like severity, occurrence, and detection.
- Efficiency and productivity: By addressing potential failure modes, FMEA can lead to more efficient and productive operations, reducing downtime and operational disruptions.
- Continuous improvement: FMEA fosters an approach of constant enhancement by encouraging organizations to examine and update their analyses on a frequent basis when new information becomes available or as products or processes change.

Disadvantages of FMEA:

- Resource intensive: Performing FMEA requires time, expertise, and resources, especially for complex systems or processes. Small organizations with limited resources may find it challenging to allocate the necessary personnel for FMEA.
- Subjectivity: The assessment of severity, occurrence, and detection ratings in FMEA often relies on expert judgment, which can introduce subjectivity into the analysis. Different teams or individuals may assign different ratings.
- Complexity: In complex systems, FMEA can become intricate and time-consuming. Managing and analyzing a large number of failure modes and their interactions can be challenging.
- Limited predictive power: FMEA is based on historical data and expert judgment, which may not always accurately predict future failure modes, especially for innovative or unique systems.
- Potential for overemphasis on high-risk items: FMEA may lead to an overemphasis on high-risk failure modes, potentially neglecting lower-risk but still important issues.

- Lack of real-world data: FMEA relies on assumptions and available data. In some cases, real-world data may be limited or unavailable, which can affect the accuracy of assessments.
- Not a standalone solution: FMEA is most effective when integrated into a broader quality and risk management framework. Relying solely on FMEA may not address all quality and risk-related issues.

10.6.7 Job Hazard Analysis (JHA)

Job hazard analysis (JHA), also known as job safety analysis (JSA), is a systematic process used to identify, assess, and mitigate workplace hazards associated with specific job tasks or activities. The primary goal of JHA is to enhance workplace safety by preventing accidents and injuries. JHA is a proactive approach to workplace safety that helps organizations reduce the risk of accidents and injuries by systematically addressing hazards associated with specific job tasks. Regular reviews and updates of JHAs are essential to maintaining a safe work environment [12].

10.6.7.1 Importance of JHA

Conducting a JHA is a critical process in occupational safety and health management. JHAs are conducted to identify and mitigate workplace hazards, thereby enhancing safety and reducing the risk of accidents and injuries. It not only protects employees from harm but also benefits organizations by reducing costs, improving productivity, and fostering a positive safety culture. Here are several key reasons why organizations should conduct a JHA:

(i) Risk identification: JHAs systematically identify potential hazards associated with specific job tasks, processes, or activities. This proactive approach helps organizations recognize and understand the risks employees face while performing their duties.

(ii) Accident prevention: By identifying and addressing hazards, JHAs help prevent workplace accidents, injuries, and illnesses. This is crucial for protecting employees' well-being and reducing the financial and human costs of accidents.

(iii) Compliance with regulations: Many regulatory agencies and occupational safety laws require employers to conduct hazard assessments and take measures to protect employees from known hazards. JHAs help organizations comply with these legal requirements.

(iv) Improved training: JHAs provide a basis for developing effective training programs. By understanding the hazards associated with their tasks, employees can be better trained on how to work safely and mitigate risks.

(v) Workplace safety culture: Promoting a culture of safety within an organization is essential for reducing accidents. JHAs contribute to this culture by encouraging employees and management to prioritize safety and collaborate on hazard identification and control.

(vi) Efficiency and productivity: Identifying and mitigating hazards can lead to more efficient work processes. When employees feel safer, they are often more focused and productive.

(vii) Reduced workers' compensation costs: Fewer workplace injuries and accidents mean reduced workers' compensation costs for employers. By preventing injuries, organizations can avoid the financial burden associated with medical expenses, lost workdays, and compensation claims.

(viii) Improved equipment and tools: JHAs may reveal issues with equipment or tools that need repair or replacement. Addressing these issues promptly can prevent accidents caused by malfunctioning equipment.

(ix) Better emergency preparedness: Understanding job hazards enables organizations to develop effective emergency response plans. Knowing the specific risks associated with each job task can improve the response to accidents or emergencies.

(x) Continuous improvement: JHAs should be a part of an ongoing safety management system. Regularly reviewing and updating JHAs as circumstances change or new hazards are identified helps organizations continuously improve safety.

(xi) Employee involvement: Involving employees in the JHA process encourages them to take ownership of safety. Employees often have valuable insights into job hazards and can contribute to solutions.

(xii) Legal defense: In the unfortunate event of a workplace accident or lawsuit, having documented JHAs can serve as evidence that an organization took reasonable steps to identify and mitigate hazards.

10.6.7.2 How to Conduct a JHA

Conducting a JHA involves a systematic process to identify and mitigate workplace hazards associated with specific job tasks or activities. JHA is an ongoing process and should be part of an organization's safety management system. Regularly reviewing and updating JHAs ensures that workplace hazards are continuously identified and mitigated, leading to a safer work environment. Here are the steps to conduct a JHA:

(i) Select the job or task for analysis: Choose a specific job or task that one want to analyze. It is often best to start with high-risk tasks or those with a history of accidents or near-misses.

(ii) Assemble a JHA team: Form a team that includes individuals with expertise in the job or task being analyzed. This may include employees who perform the job, supervisors, safety professionals, and subject matter experts.

(iii) Break down the job into steps: Divide the selected job or task into a sequence of discrete steps or activities. Each step should represent a specific action or task performed.

(iv) Identify hazards for each step: For each step in the job, systematically identify potential hazards. Hazards can be physical, chemical, biological, ergonomic, or psychosocial. Consider sources of energy, equipment, tools, materials, environment, and human factors.

(v) Determine the consequences of hazards: Assess the potential consequences or outcomes if each identified hazard were to cause an accident or incident. Consider the severity of injuries, illnesses, property damage, or environmental impact.

(vi) Determine risk factors: Evaluate the likelihood or probability of each hazard causing harm. Factors that influence risk include frequency of exposure, duration of exposure, and the effectiveness of existing controls.

(vii) Control measures: Identify existing control measures that are in place to mitigate each hazard. These may include engineering controls, administrative controls, personal protective equipment, or standard operating procedures.

(viii) Evaluate the effectiveness of controls: Assess how well the existing control measures are working to eliminate or reduce the hazards. Determine if there are gaps in control measures that need improvement.

(ix) Determine risk levels: Calculate the risk level for each hazard by considering the severity and likelihood of each hazard. This can be done using a risk matrix or a numerical rating system.

(x) Prioritize hazards: Prioritize hazards based on their risk levels. Focus on high-risk hazards that have the potential for severe consequences and those with a higher likelihood of occurrence.

(xi) Develop hazard controls: Identify and implement control measures to mitigate or eliminate high-priority hazards. Controls should be feasible, effective, and tailored to each specific hazard.

(xii) Document the JHA: Create a written JHA document that includes all the information gathered during the analysis, including identified hazards, consequences, controls, risk levels, and recommended actions.

(xiii) Review and verify: Have the JHA reviewed by the JHA team, supervisors, and affected employees to ensure accuracy and completeness. Verify that the recommended control measures are practical and can be implemented.

(xiv) Implement controls: Put the recommended control measures into practice. Ensure that employees are trained on these measures and understand their importance.

(xv) Monitor and review: Continuously monitor the job or task to ensure that the implemented controls are effective. Periodically review and update the JHA as circumstances change or new hazards are identified.

(xvi) Communicate and train: Communicate the findings of the JHA to all employees involved in the job or task. Provide training on safe work practices and the use of control measures.

(xvii) Feedback and improvement: Encourage employees to provide feedback on the effectiveness of controls and the JHA process. Use this feedback for continuous improvement.

10.6.7.3 Advantages and Disadvantages of JHA

JHA is a systematic approach to identifying and mitigating workplace hazards. Like any safety management tool, JHA has its advantages and disadvantages. When performed effectively and integrated into a comprehensive safety program, JHA can significantly enhance workplace safety, reduce accidents, and improve overall organizational efficiency. Here is a breakdown of these pros and cons:

Advantages of JHA:

- Hazard identification: JHA helps identify potential workplace hazards systematically. It encourages a thorough examination of each job task, ensuring that hazards are not overlooked.
- Injury prevention: By identifying and addressing hazards, JHA helps prevent workplace injuries and accidents. This is vital for employee safety and well-being.
- Increased awareness: JHA raises awareness about safety among employees and management. It fosters a safety-conscious culture where everyone actively participates in identifying and mitigating hazards.
- Compliance with regulations: Many regulatory agencies require employers to assess and control workplace hazards. JHA helps organizations meet these legal requirements and avoid fines or penalties.
- Training enhancement: JHA provides a foundation for designing effective safety training programs. Employees can be trained to recognize and mitigate the hazards associated with their specific job tasks.
- Efficiency improvement: By addressing hazards, organizations can optimize work processes, leading to increased efficiency and productivity.
- Cost reduction: Fewer workplace accidents mean reduced workers' compensation costs and medical expenses. JHA can lead to cost savings for organizations.

- Continuous improvement: JHA is an ongoing process. Regular reviews and updates ensure that hazards are managed effectively, contributing to continuous improvement in workplace safety.

Disadvantages of JHA:

- Time and resource intensive: Conducting a JHA can be time-consuming, especially in complex workplaces with numerous job tasks. It requires the commitment of resources, including personnel and time.
- Subjectivity: JHA often relies on the judgment and experience of individuals conducting the analysis. This subjectivity can lead to variations in hazard identification and assessment.
- Resistance to change: Some employees and organizations may resist the changes or additional safety measures recommended by JHA, perceiving them as cumbersome or disruptive.
- Complexity: In environments with constantly changing conditions or multiple interacting hazards, conducting a comprehensive JHA can be complex and challenging.
- Overemphasis on low-risk tasks: In an effort to simplify the process, there is a risk of overemphasizing low-risk tasks and overlooking critical high-risk activities.
- Documentation burden: Proper documentation of JHA findings and action plans is essential but can be burdensome if not managed efficiently.
- Complacency: Organizations that rely solely on JHA may become complacent about safety. Completing a JHA does not guarantee safety; it is just one part of a broader safety management system.
- Limited predictive power: JHA is based on historical data and expert judgment. It may not always accurately predict future hazards, especially in innovative or unique work environments.

10.6.8 Incident and Near-Miss Analysis

Incident and near-miss analysis is a crucial component of a proactive approach to safety and risk management in various industries. This process involves the systematic examination of incidents that have occurred, as well as near-miss events, to identify root causes, lessons learned, and opportunities for improvement. In summary, incident and near-miss analysis are essential components of a robust safety and risk management program. By thoroughly investigating incidents and near-misses, organizations can identify opportunities for improvement, implement corrective actions, and ultimately enhance workplace safety and operational efficiency [13].

10.6.8.1 Incident Analysis

Incident analysis is a systematic process used to investigate and understand various types of incidents that occur in the workplace, whether they result in injuries, accidents, property damage, or operational disruptions. The primary goal of incident analysis is to determine the root causes of the incident, identify contributing factors, and develop corrective actions to prevent similar incidents from happening in the future. By conducting thorough incident analyses and addressing the underlying causes, organizations can improve workplace safety, reduce risks, and prevent accidents and injuries. It is an essential component of a comprehensive safety management program. Here is an overview of the incident analysis process:

(i) Reporting and initial response: Incidents should be reported as soon as they occur or are discovered. Prompt reporting ensures that investigations can begin promptly. An initial response may involve securing the scene, providing first aid or medical attention, and ensuring the safety of all personnel.

(ii) Gathering information: Investigative teams, often consisting of safety experts, supervisors, or relevant personnel, are assembled to gather information about the incident. Witnesses, affected employees, and others with relevant knowledge are interviewed to collect facts and details about the incident.

(iii) Preservation of evidence: Physical evidence, such as equipment, tools, or materials involved in the incident, is preserved to aid in the investigation. Photographs, sketches, and documentation can be valuable.

(iv) Root cause analysis: A critical step in incident analysis is identifying the root causes of the incident. Root causes are the fundamental reasons or systemic factors that led to the incident.

(v) Determining contributing factors: In addition to identifying root causes, investigators also determine contributing factors or secondary causes that may have played a role in the incident. Contributing factors could include human errors, equipment failures, communication breakdowns, lack of training, or inadequate safety procedures.

(vi) Report and documentation: The findings of the incident analysis are documented in an incident report. This report includes information on what happened, why it happened, and what corrective actions will be taken. The report should be clear, concise, and easily understandable to ensure that corrective actions are effectively implemented.

(vii) Corrective actions: Based on the analysis findings, corrective actions are developed to address the root causes and contributing factors. Corrective actions may include process improvements, safety procedure updates, employee training, equipment maintenance, or other measures designed to prevent similar incidents.

(viii) Implementation and monitoring: Corrective actions are implemented, and their effectiveness is monitored over time. Ongoing monitoring ensures that the changes are effective in preventing similar incidents.

(ix) Follow-up and communication: Communication of the analysis findings and corrective actions is crucial to ensure that all relevant parties are aware of the incident and the steps taken to prevent recurrence.

(x) Continuous improvement: Incident analysis is an ongoing process. Organizations should continuously review and update their safety procedures and practices based on the lessons learned from incident analysis.

10.6.8.2 Importance of Incident Analysis

Incident analysis is critical for preventing future incidents, enhancing safety, ensuring compliance, engaging employees, and supporting the overall success and sustainability of organizations. It is a proactive and data-driven approach to managing risks and improving workplace safety. Incident analysis is of paramount importance in various industries and workplaces for several key reasons:

- Prevention of recurrence: One of the primary goals of incident analysis is to identify the root causes of incidents and accidents. By understanding why an incident occurred, organizations can implement corrective actions to address the underlying issues. This helps prevent the recurrence of similar incidents, reducing the risk to employees, property, and operations.
- Enhanced safety: Incident analysis contributes to improved workplace safety. It allows organizations to identify hazards and vulnerabilities in their processes, systems, and equipment. Addressing these issues proactively enhances safety measures and reduces the likelihood of accidents and injuries.
- Continuous improvement: Incident analysis is part of a broader culture of continuous improvement. Organizations that invest in analyzing incidents and near-misses are committed to learning from their mistakes and striving for better safety performance.
- Risk mitigation: By identifying the causes of incidents, organizations can assess and mitigate risks more effectively. This proactive approach minimizes the potential impact of future incidents on personnel, the environment, and the organization's reputation.
- Legal and regulatory compliance: Many industries and regions have regulations and legal requirements that mandate the investigation and reporting of certain types of incidents. Compliance with these regulations is essential to avoid fines, penalties, and legal liabilities.
- Employee engagement: Involving employees in incident analysis fosters a culture of safety and engagement. Employees who feel their concerns are taken seriously and see corrective actions implemented are more likely to be proactive in reporting hazards and near-misses.
- Insurance and liability management: Incident analysis provides organizations with a clear understanding of the factors contributing to incidents. This information can be valuable when working with insurance providers and managing liability claims.
- Operational efficiency: By identifying and addressing factors that contribute to incidents, organizations can improve the

efficiency of their processes and operations. This can lead to cost savings and operational improvements.

- Reputation management: Incidents can have a significant impact on an organization's reputation, especially if they involve harm to employees, customers, or the environment. Effective incident analysis and response can help protect an organization's reputation and brand.
- Data-driven decision-making: Incident data collected and analyzed over time can provide valuable insights for data-driven decision-making. Organizations can use historical incident data to identify trends, assess the effectiveness of safety measures, and allocate resources where they are needed most.
- Resource allocation: Incident analysis helps organizations allocate resources more effectively. By understanding which areas or processes have a higher incidence of incidents, organizations can prioritize safety initiatives and resource allocation accordingly.

10.6.8.3 Steps Involved in Incident Analysis

Incident analysis is a systematic process used to investigate and understand the causes and contributing factors of an incident, whether it is a workplace accident, near-miss, or other unexpected event. The goal of incident analysis is to identify the root causes and lessons learned to prevent similar incidents in the future. Organizations can conduct thorough incident analyses, identify root causes, and implement effective corrective actions to prevent future incidents and enhance workplace safety. Here are the typical steps involved in incident analysis:

(i) Immediate response and reporting: Ensure that immediate medical attention is provided to injured parties if necessary. Secure the scene to prevent further harm or damage. Report the incident to the appropriate individuals or authorities, including safety officers, supervisors, or regulatory agencies, as required.

(ii) Form an investigation team: Assemble a team of qualified individuals to conduct the investigation. This team should include personnel with relevant expertise, such as safety professionals, subject matter experts, and affected employees.

(iii) Preserve evidence: Secure and preserve all evidence related to the incident, including physical evidence, equipment, records, and witness statements. This helps ensure the accuracy of the investigation.

(iv) Define the scope: Clearly define the scope and objectives of the investigation. Identify the specific incident or problem that needs analysis and determine the boundaries of the investigation.

(v) Gather information: Collect as much information as possible about the incident. This includes witness statements, photographs, videos, documents, equipment logs, and any other relevant data.

(vi) Sequence of events: Create a timeline of events leading up to and following the incident. This sequence should include all activities, actions, and conditions that may have contributed to the incident.

(vii) Identify root causes: Use systematic root cause analysis methods to identify the underlying or root causes of the incident. Consider factors related to human behavior, equipment, processes, environmental conditions, and organizational culture.

(viii) Contributing factors: Identify and analyze contributing factors or intermediate causes that may have played a role in the incident. These are factors that may not be root causes themselves but contributed to the incident.

(ix) Recommendations and corrective actions: Develop a list of recommendations and corrective actions to address the identified root causes and contributing factors. These actions should be practical, specific, and aimed at preventing similar incidents. Prioritize these actions based on their potential to reduce risk and enhance safety.

(x) Report and documentation: Prepare a comprehensive incident report that summarizes the investigation process, findings, root causes, contributing factors, and recommended actions. Ensure that the report is clear, concise, and well-documented for future reference.

(xi) Communication: Communicate the findings and recommendations to all relevant stakeholders, including management, employees, and regulatory authorities, as appropriate.

(xii) Implementation of corrective actions: Execute the recommended corrective actions promptly and effectively. Assign responsibility for each action, establish timelines, and monitor progress.

(xiii) Follow-up and monitoring: Continuously monitor the progress of corrective actions to ensure they are implemented as intended and that they effectively address the identified issues. Review and evaluate the effectiveness of the corrective actions over time.

(xiv) Lessons learned: Encourage organizational learning by discussing the incident analysis findings and lessons learned with employees and teams. Use this knowledge to improve safety policies, procedures, and training programs.

(xv) Documentation and records retention: Maintain detailed records of the incident analysis process, including all documents, reports, and communication related to the incident.

10.6.8.4 Advantages and Disadvantages of Incident Analysis

Incident analysis, which involves the systematic examination of incidents that have occurred in the workplace or other settings, serves as a valuable tool for learning, improvement, and risk management. Organizations can mitigate some of the challenges by fostering a culture of open reporting, providing training in incident analysis, and ensuring that investigations are conducted impartially and with a focus on prevention rather than blame. Like any process, it comes with its own set of advantages and disadvantages:

Advantages of Incident Analysis:

- Learning opportunity: Incident analysis provides a valuable learning opportunity for organizations. It allows them to gain insights into what went wrong, why it happened, and how to prevent similar incidents in the future.
- Continuous improvement: By identifying the root causes of incidents, organizations can implement corrective and preventive actions to improve safety, operational efficiency, and overall performance.
- Risk reduction: Understanding the causes of incidents enables organizations to proactively mitigate risks and

hazards, reducing the likelihood of future incidents and associated costs.

- Enhanced safety culture: An incident analysis process that encourages reporting and investigation fosters a culture of safety. Employees become more aware of potential risks and hazards and are more likely to report incidents and near-miss events.
- Regulatory compliance: In many industries, incident analysis is a regulatory requirement. Conducting thorough investigations and reporting incidents helps organizations comply with legal and regulatory obligations.
- Improved documentation: Incident analysis produces documentation that can serve as a historical record. It helps organizations track trends, monitor progress, and demonstrate their commitment to safety.

Disadvantages of Incident Analysis:

- Resource intensive: Incident analysis can be time-consuming and resource-intensive. It requires skilled personnel to conduct investigations, collect data, and develop corrective actions.
- Subjectivity: The quality of incident analysis depends on the objectivity and expertise of the investigators. Bias or incomplete information can lead to inaccurate conclusions.
- Blame culture: In some organizations, incident analysis may be perceived as a way to assign blame or fault rather than as a means to identify root causes and prevent future incidents. This can create a culture of fear and hinder reporting.
- Complexity: Incidents are often multifaceted, with multiple contributing factors. Analyzing these complex scenarios and determining root causes can be challenging.
- Resistance to change: Implementing corrective actions based on incident analysis findings may face resistance, especially if changes disrupt established routines or practices.
- Confidentiality concerns: In some cases, incidents involve sensitive information that organizations may be hesitant to disclose publicly or internally. Balancing transparency with privacy can be a challenge.
- Limited predictive power: While incident analysis is valuable for understanding past incidents, it may have limited

predictive power for future events, especially in unique or unpredictable circumstances.

- Reporting barriers: Employees may be reluctant to report incidents due to fear of reprisal or a belief that reporting is futile. This can hinder the effectiveness of incident analysis.

10.6.8.5 Near-Miss Analysis

Near-miss analysis is a crucial component of proactive safety management in various industries. It involves the systematic examination and analysis of near-miss events, which are incidents that did not result in harm or damage but had the potential to do so. The purpose of near-miss analysis is to identify underlying causes, vulnerabilities, and lessons learned to prevent future incidents and improve safety. Near-miss analysis helps organizations proactively identify and address potential hazards and vulnerabilities, ultimately enhancing safety, reducing risks, and preventing more serious incidents.

10.6.8.6 Importance of Near-Miss Analysis

Near-miss analysis holds significant importance in safety and risk management across various industries. Near-miss analysis is essential for proactive risk management, accident prevention, and the cultivation of a safety-focused culture within organizations. By addressing potential hazards and vulnerabilities early on, organizations can reduce risks, save costs, and create safer work environments for their employees. Here are key reasons why near-miss analysis is essential:

- Accident prevention: Near-miss events are often early indicators of potential hazards and weaknesses in safety systems. Analyzing these events allows organizations to identify and address these issues before they lead to actual accidents, injuries, or damage.
- Proactive risk management: Near-miss analysis encourages organizations to take a proactive approach to risk management. It helps prevent incidents by identifying and mitigating underlying causes and contributing factors.
- Enhanced safety culture: Promoting the reporting and analysis of near-miss events fosters a safety-conscious culture within an organization. Employees become more aware of

potential risks and are encouraged to actively participate in safety initiatives.

- Learning opportunities: Near-miss events provide valuable learning opportunities. Analyzing these events can reveal insights into the effectiveness of safety procedures, training programs, and hazard controls.
- Data for improvement: The data collected from near-miss analysis can inform data-driven decision-making. Organizations can use this data to improve safety policies, procedures, and equipment.
- Cost savings: Preventing accidents and incidents through near-miss analysis leads to cost savings. It reduces expenses related to medical treatment, workers' compensation, property damage, and legal liabilities.
- Compliance and regulation: In many industries, regulatory agencies require organizations to assess and address workplace hazards. Near-miss analysis helps organizations comply with these legal requirements and demonstrates a commitment to safety.
- Increased employee engagement: Involving employees in near-miss reporting and analysis empowers them to contribute to safety improvement. It strengthens their sense of ownership in maintaining a safe work environment.
- Continuous improvement: Near-miss analysis contributes to continuous improvement in safety performance. It allows organizations to regularly evaluate and update safety measures based on real-world incidents and lessons learned.
- Early warning system: Near-miss analysis serves as an early warning system for identifying potential risks and vulnerabilities in processes, equipment, and systems. Addressing these issues before they escalate is more cost-effective and less disruptive.
- Improved incident response: Lessons learned from near-miss analysis can improve the organization's preparedness and response to actual incidents and emergencies.
- Competitive advantage: Organizations known for their strong safety records and proactive safety measures often have a competitive advantage. They attract and retain employees and customers who prioritize safety.

- Organizational reputation: Demonstrating a commitment to safety through near-miss analysis enhances an organization's reputation, both internally and externally.

10.6.8.7 Steps for Conducting Near-Miss Analysis

Conducting a thorough near-miss analysis is essential for identifying potential hazards and improving safety in the workplace. Near-miss events provide valuable opportunities to address issues before they escalate into accidents or injuries. Organizations can effectively analyze near-miss events, identify and address root causes, and proactively improve safety in the workplace. Near-miss analysis is a valuable tool for preventing accidents, enhancing safety culture, and reducing risks. Here are the typical steps involved in near-miss analysis:

(i) Near-miss reporting: Encourage and establish a culture of near-miss reporting where employees feel comfortable reporting incidents without fear of reprisal. Ensure that there are clear and accessible reporting mechanisms in place, such as an incident reporting system or designated personnel to receive reports.

(ii) Immediate response and scene preservation: If necessary, provide immediate assistance to anyone involved in the near-miss event, ensuring their safety and well-being. Preserve the scene and any evidence related to the near-miss event to maintain the integrity of the investigation.

(iii) Form an investigation team: Assemble a team of qualified individuals, including safety professionals, subject matter experts, and relevant employees, to conduct the near-miss analysis.

(iv) Documentation and data collection: Gather all available information about the near-miss event, including witness statements, photographs, videos, records, and any other relevant data. Create a timeline of events leading up to and following the near-miss incident.

(v) Root cause analysis: Apply root cause analysis methods to identify the root causes of the near-miss event. Explore why the near-miss occurred and uncover the underlying factors contributing to it.

(vi) Contributing factors: Identify and analyze contributing factors or intermediate causes that played a role in the near-miss event. These factors may not be root causes themselves but contributed to the event's potential.

(vii) Recommendations and corrective actions: Develop recommendations based on the findings of the near-miss analysis. These recommendations should focus on preventing similar incidents in the future. Specify corrective actions that address the root causes and contributing factors, making them practical, specific, and actionable.

(viii) Report and documentation: Prepare a comprehensive near-miss analysis report summarizing the investigation process, findings, root causes, contributing factors, recommendations, and corrective actions. Ensure that the report is clear, concise, and well-documented for future reference.

(ix) Communication: Communicate the findings, recommendations, and corrective actions to all relevant stakeholders, including management, employees, and safety committees. Promote transparency and encourage feedback and input from employees.

(x) Implementation of corrective actions: Execute the recommended corrective actions promptly and effectively. Assign responsibility for each action, establish timelines, and monitor progress.

(xi) Follow-up and monitoring: Continuously monitor the progress of corrective actions to ensure they are implemented as intended and effectively address the identified issues. Review and evaluate the effectiveness of the corrective actions over time.

(xii) Lessons learned: Share the findings and lessons learned from the near-miss analysis with employees and teams to promote organizational learning. Use this knowledge to improve safety policies, procedures, training programs, and risk management practices.

(xiii) Continuous improvement: Integrate near-miss analysis into the ongoing safety management system to continuously identify and address potential hazards and vulnerabilities. Regularly review and update the near-miss analysis process based on feedback and evolving safety needs.

(xiv) Documentation and records retention: Maintain detailed records of the near-miss analysis process, including all documents, reports, and communication related to the near-miss events.

10.6.8.8 Advantages and Disadvantages of Near-Miss Analysis

Near-miss analysis is a proactive safety management practice that involves examining incidents or situations that had the potential to cause harm but resulted in no injuries or damage. Despite these potential disadvantages, the benefits of near-miss analysis in preventing accidents, improving safety culture, and reducing costs far outweigh the drawbacks. Organizations can address some of these challenges by implementing clear reporting procedures, providing training on near-miss analysis, and fostering a culture that values safety and learning from near-miss events. Like any safety tool, it has its advantages and disadvantages:

Advantages of Near-Miss Analysis:

- Prevention of serious incidents: By identifying and analyzing near-miss events, organizations can proactively address underlying issues and prevent future accidents that could have more severe consequences.
- Improved safety culture: Encouraging near-miss reporting fosters a culture of safety where employees actively contribute to identifying and mitigating hazards, increasing overall safety awareness.
- Cost savings: Preventing accidents and injuries through near-miss analysis can lead to significant cost savings, including reduced workers' compensation claims, medical expenses, and property damage costs.
- Learning opportunities: Near-miss events provide valuable learning opportunities for organizations. They help identify weaknesses in safety procedures, equipment, and training programs.
- Proactive risk reduction: Near-miss analysis allows organizations to address risks and vulnerabilities before they lead to actual harm, reducing the likelihood of future incidents.
- Data for continuous improvement: Data from near-miss reports can be used to refine safety policies, procedures,

and training programs, contributing to ongoing safety improvement.

- Increased employee engagement: Involving employees in the near-miss reporting and analysis process empowers them to take an active role in safety management.
- Legal and regulatory compliance: Identifying and addressing near-miss events can help organizations comply with safety regulations and demonstrate a commitment to safety to regulatory agencies.

Disadvantages of Near-Miss Analysis:

- Underreporting: Employees may be reluctant to report near-miss events due to fear of reprisal, lack of awareness, or perceptions that reporting is time-consuming.
- Subjectivity: The effectiveness of near-miss analysis can be influenced by the subjectivity of investigators and analysts. Different individuals may interpret near-miss events differently.
- Resource intensive: Conducting thorough near-miss analyses can be resource-intensive, requiring time, personnel, and documentation efforts.
- Lack of standardization: Organizations may lack standardized processes for near-miss reporting and analysis, leading to inconsistencies in how events are handled.
- Overemphasis on reporting: Organizations may become overly focused on reporting near-miss events at the expense of addressing underlying safety issues or implementing corrective actions.
- Limited predictive value: While near-miss analysis is valuable for identifying potential hazards, it may not always accurately predict future incidents, especially when conditions change.
- Privacy concerns: Employees may have concerns about the privacy and confidentiality of near-miss reports, especially if they fear potential consequences for reporting.
- Resistance to change: Implementing recommended corrective actions based on near-miss analyses may face resistance from employees or management who are resistant to change.

References

1. Ozog, H., Hazard identification, analysis and control. *Chem. Eng. (NY), (United States)*, 92, 4, 1985.
2. Domínguez, C.R., Mares, J.I., Hernández, R.G., Hazard identification and analysis in work areas within the Manufacturing Sector through the HAZID methodology. *Process Saf. Environ. Prot.*, 145, 23–38, 2021.
3. Cameron, I., Mannan, S., Németh, E., Park, S., Pasman, H., Rogers, W., Seligmann, B., Process hazard analysis, hazard identification and scenario definition: Are the conventional tools sufficient, or should and can we do much better? *Process Saf. Environ. Prot.*, 110, 53–70, 2017.
4. Crawley, F. and Tyler, B., *HAZOP: Guide to best practice*, Elsevier, Netherlands, 2015.
5. Rossing, N.L., Lind, M., Jensen, N., Jørgensen, S.B., A functional HAZOP methodology. *Comput. Chem. Eng.*, 34, 2, 244–253, 2010.
6. Dunjó, J., Fthenakis, V., Vílchez, J.A., Arnaldos, J., Hazard and operability (HAZOP) analysis. A literature review. *J. Hazard. Mater.*, 173, 1-3, 19–32, 2010.
7. Matsika, E., Ricci, S., Mortimer, P., Georgiev, N., O'Neill, C., Rail vehicles, environment, safety and security. *Res. Transp. Econ.*, 41, 1, 43–58, 2013.
8. Čepin, M. and Čepin, M., Event tree analysis, in: *Assessment of Power System Reliability: Methods and Applications*, pp. 89–99, 2011.
9. Huber, T.E., Preliminary safety analysis report for project 89-GEB-610 Plutonium Finishing Plant instrumentation upgrade. Revision 1, Westinghouse Hanford Co, 1995.
10. Fiorentini, L., *Bow-Tie Industrial Risk Management Across Sectors: A Barrier-Based Approach*, John Wiley & Sons, United States, 2021.
11. Mikulak, R.J., McDermott, R., Beauregard, M., *The basics of FMEA*, CRC press, New York, 2017.
12. Channing, J. (Ed.), *Safety at work*, Routledge, London, 2013.
13. Dulac, N., A framework for dynamic safety and risk management modeling in complex engineering systems (Doctoral dissertation), Massachusetts Institute of Technology, 2007.

11

Safety Function Deployment and Quantification of Basic Events

Abstract

This chapter describes the safety function deployment and quantification of basic events, how they ensure safety during a product or system's design and operation, and how they can be used together to enhance safety assessments and decision-making. In the latter part, a detailed discussion was provided about their significance, important steps to conduct analysis, and their advantages and disadvantages.

***Keywords*:** Safety function deployment, probabilistic risk assessment, quantification of basic events

11.1 Background

Safety function deployment (SFD) and the quantification of basic events are two complementary approaches used in the field of safety and risk management to ensure that safety considerations are systematically integrated into a product or system's design and operation. While they serve different purposes, they can be used together to enhance safety assessments and decision-making. By integrating SFD and the quantification of basic events, organizations can systematically address safety concerns, make informed decisions about risk management, and enhance the overall safety of their products or systems.

11.2 Introduction

Safety function deployment is a structured methodology that focuses on identifying, defining, and integrating safety functions or features into a product or system. It is primarily concerned with the design and

Abhishek Barua, Kanchan Kumari, Sumit Saha, Siddharth Jeet and Swastik Pradhan. Institutional and Industrial Safety Engineering Practices, (267–286) © 2025 Scrivener Publishing LLC

implementation of safety measures to prevent or mitigate accidents, incidents, or adverse events. Quantification of basic events is a component of probabilistic risk assessment (PRA), and it involves assigning probabilities and consequences to the lowest-level events or conditions (basic events) that can contribute to undesired outcomes or system failures. Both SFD and quantification of basic events are essential tools for managing safety and risk in complex systems and product development. Whereas SFD focuses on integrating safety features into the design process, quantification of basic events, as part of PRA, provides a quantitative framework for assessing and prioritizing risks associated with those safety functions. Together, they contribute to safer and more reliable products and systems across various industries [1].

11.3 Safety Function Deployment (SFD)

Safety function deployment (SFD) is a systematic and structured approach used in product design and development to ensure that safety considerations are integrated throughout the entire process. SFD is closely related to quality function deployment, a methodology used for translating customer requirements into product or service design characteristics. SFD, however, focuses specifically on safety aspects. Safety function deployment is especially important in industries where safety is paramount, such as automotive manufacturing (for designing safety features in vehicles), medical device manufacturing, aerospace, and chemical processing. By systematically integrating safety considerations from the earliest stages of product development, SFD helps minimize the risks associated with products and systems, ensuring the safety of users and the environment [2].

11.3.1 Important Parts of Safety Function Deployment

Safety function deployment (SFD) is a structured approach used in various industries, particularly in product design and engineering, to integrate safety considerations into the development process. It helps ensure that safety requirements and features are systematically incorporated into a product or system. SFD consists of several parts or stages, each with its own set of activities. Organizations can reduce the likelihood of safety-related incidents and ensure that their products meet safety standards and regulations. Here are the key parts of SFD:

a) Project Initiation:
   - Define Objectives: Clearly articulate the goals and objectives of the SFD process, including the specific safety-related outcomes that need to be achieved.
   - Identify Stakeholders: Identify all relevant stakeholders, including designers, engineers, safety experts, and end-users, who will be involved in the process.

b) Safety Requirement Identification:
   - Gather Information: Collect data and information about safety standards, regulations, and best practices applicable to the product or system being developed.
   - Identify Safety Requirements: Determine the safety requirements that must be met to ensure the safe operation of the product or system. These requirements may include hazard analysis, risk assessments, and safety goals.

c) Safety Function Definition:
   - Translate Requirements: Translate safety requirements into specific safety functions or features that need to be integrated into the product or system.
   - Prioritize Functions: Prioritize safety functions based on their criticality and potential impact on safety.

d) Design Integration:
   - Incorporate Functions: Integrate the identified safety functions into the overall design of the product or system. This involves working closely with design and engineering teams to ensure that safety features are seamlessly integrated.
   - Assess Impact: Evaluate the impact of safety functions on the product's performance, functionality, and cost.

e) Validation and Verification:
   - Test and Validate: Conduct tests and simulations to verify that the safety functions are effective in mitigating identified hazards and risks.
   - Iterate as Needed: If issues or shortcomings are identified during validation, then make necessary iterations to improve safety functions.

f) Documentation and Compliance:
   - Documentation: Maintain thorough documentation of the SFD process, including safety requirements, design

decisions, test results, and compliance with safety standards and regulations.
- Regulatory Compliance: Ensure that the product or system complies with relevant safety regulations and standards.

g) Communication and Training:
- Communication: Keep all stakeholders informed about the progress of SFD and any safety-related developments.
- Training: Provide training and education to relevant personnel on the use and maintenance of safety functions and features.

h) Continuous Improvement:
- Feedback Loop: Establish a feedback loop for continuous improvement. Monitor the performance of safety functions in real-world scenarios and make enhancements or updates as needed.
- Lessons Learned: Capture lessons learned from the SFD process to inform future projects and improve safety practices.

11.3.2 Importance of Safety Function Deployment

Safety function deployment is important because it helps organizations systematically integrate safety considerations into their products, systems, or processes. This methodology is a crucial component of safety engineering and risk management. This proactive approach not only enhances safety but also contributes to overall quality, compliance, cost savings, and customer satisfaction. It is a fundamental practice in industries where safety is paramount, such as aerospace, automotive, healthcare, and manufacturing. Here are some of the key reasons why SFD is important:

- Proactive Safety Planning: SFD encourages a proactive approach to safety by incorporating safety requirements and considerations into the early stages of product or process development. This helps prevent safety issues and hazards from arising later in the design or operational phases, reducing the need for costly retrofits and redesigns.
- Risk Reduction: By identifying and addressing potential safety risks and hazards early in the development process, SFD helps reduce the likelihood of accidents, incidents,

injuries, and damage. This, in turn, minimizes associated costs and liabilities.

- Compliance with Regulations and Standards: Many industries are subject to strict safety regulations and standards. Implementing SFD ensures that safety requirements are met, helping organizations comply with legal and regulatory obligations. Non-compliance can lead to legal and financial consequences.
- Enhanced Product Quality: Integrating safety considerations into the design and development process can lead to improved product quality. Safety features can also enhance the overall functionality, reliability, and usability of products, making them more competitive in the market.
- Customer Satisfaction: Safety is a significant concern for customers and end-users. Products or processes that prioritize safety are more likely to gain customer trust and satisfaction. SFD helps in identifying and addressing safety-related customer concerns.
- Cost Savings: Addressing safety issues early in the development process is typically less expensive than dealing with safety-related problems after a product is in use. SFD helps organizations avoid costly recalls, lawsuits, and reputation damage.
- Integrated Approach: SFD promotes collaboration among various stakeholders, including engineers, safety experts, designers, and management. This integrated approach ensures that safety considerations are not isolated but are considered holistically within the broader context of the project.
- Continuous Improvement: SFD is an iterative process that encourages ongoing safety assessment and improvement. It allows organizations to learn from past safety incidents or near-misses and make necessary adjustments to enhance safety measures continually.
- Risk Communication: SFD facilitates effective communication of safety-related information within an organization. It helps in conveying safety requirements, goals, and priorities to all team members, ensuring a shared understanding of safety objectives.

- Competitive Advantage: Demonstrating a commitment to safety through SFD can be a source of competitive advantage. It can differentiate a company's products or processes in the market, attracting safety-conscious customers and partners.

11.3.3 Important Steps Involved in Safety Function Deployment

Safety function deployment is a structured methodology used to embed safety into every stage of a product's lifecycle, ultimately reducing the likelihood of accidents, injuries, and product recalls while enhancing customer trust and satisfaction. It helps organizations ensure that safety is a fundamental aspect of their products or processes from the outset. Here are the important steps involved in SFD:

a) Define Safety Objectives: The process begins by clearly defining safety objectives and goals. These objectives should be specific, measurable, achievable, relevant, and time-bound (SMART). They serve as the foundation for the entire SFD process.
b) Identify Stakeholders: Identify all stakeholders who have an interest in the safety of the product or process. This includes end-users, regulatory bodies, internal teams, and anyone else who may be affected by safety-related issues.
c) Create a Safety Requirement List: Develop a comprehensive list of safety requirements based on the safety objectives and stakeholder input. These requirements should cover various aspects of safety, including functional safety, product liability, and compliance with safety standards and regulations.
d) Risk Assessment: Perform a thorough risk assessment to identify potential hazards and risks associated with the product or process. Use techniques like hazard and operability studies and failure modes and effects analysis to identify and prioritize risks.
e) Link Safety Requirements to Risks: Connect each safety requirement to the specific risks it addresses. This linkage helps ensure that safety requirements are driven by identified risks and hazards.

f) Prioritize Safety Requirements: Prioritize safety requirements based on the severity of associated risks and the likelihood of those risks occurring. High-risk requirements should receive greater attention and resources.
g) Design Safety Functions: Develop safety functions and features that address the identified safety requirements. These functions should be designed to prevent, detect, or mitigate safety risks effectively.
h) Implement Safety Measures: Integrate safety measures into the product or process design. This may involve incorporating safety sensors, alarms, fail-safes, interlocks, and other safety features.
i) Verification and Validation: Verify and validate the safety functions and measures through testing, simulation, and analysis. Ensure that they meet the specified safety requirements and effectively address the identified risks.
j) Documentation: Maintain comprehensive documentation of all safety-related decisions, requirements, design choices, and verification/validation results. This documentation is crucial for compliance, auditing, and future reference.
k) Continuous Improvement: SFD is an iterative process. Continuously monitor safety performance, gather feedback from users, and update safety requirements and functions as needed to improve safety over time.
l) Communication and Training: Communicate safety-related information and requirements to all relevant stakeholders, including end-users, suppliers, and internal teams. Provide training as necessary to ensure that safety procedures and measures are understood and followed.
m) Compliance and Certification: Ensure that the product or process complies with relevant safety standards and regulations. Obtain necessary certifications and approvals from regulatory bodies as required.

11.3.4 Advantages and Disadvantages of Safety Function Deployment

Safety function deployment can be a valuable tool for integrating safety into the design and development process, but it comes with its own set of challenges and considerations. Organizations need to weigh the advantages and disadvantages to determine whether SFD is a suitable approach for their specific needs and contexts. Like any approach, it has its advantages and disadvantages.

Advantages of SFD:

- Holistic Safety Integration: SFD encourages a comprehensive approach to safety, ensuring that safety considerations are integrated from the early stages of product or system development. This can lead to more effective safety measures and fewer safety-related issues down the line.
- Risk Reduction: By systematically identifying potential safety hazards and their causes, SFD helps in reducing the risks associated with a product or system. This can lead to fewer accidents, injuries, and safety incidents.
- Improved Communication: SFD facilitates communication between different stakeholders involved in the design and development process, including engineers, safety experts, and management. It ensures that everyone is on the same page regarding safety requirements and objectives.
- Customer Satisfaction: Addressing safety concerns proactively through SFD can enhance customer satisfaction. Consumers are more likely to trust and use products or systems that are designed with their safety in mind.
- Legal and Regulatory Compliance: SFD helps organizations comply with safety regulations and standards, reducing the risk of legal liabilities and penalties.
- Cost Savings: Detecting safety issues early in the design phase is typically less expensive than addressing them later in the production or operational phases. SFD can save costs associated with redesign and retrofits.

Disadvantages of SFD:

- Complexity: SFD can be a complex process, especially for large and intricate systems. It requires significant time and expertise to implement properly.
- Resource Intensive: Properly conducting SFD may require dedicated safety experts and resources, which can be costly for organizations, especially smaller ones.

- Resistance to Change: In some cases, there may be resistance to integrating safety into established design processes, as it may disrupt existing workflows and timelines.
- Subjectivity: Assessing safety requirements and priorities can be subjective and dependent on the judgment of the individuals involved. This subjectivity can lead to variations in safety outcomes.
- Potential for Over-Engineering: In an effort to maximize safety, there is a risk of over-engineering a product or system, which can lead to increased costs and complexity without a corresponding increase in safety benefits.
- Limited Effectiveness Without Commitment: SFD is most effective when there is a strong commitment to safety at all levels of an organization. Without this commitment, it may not lead to significant safety improvements.

11.4 Probabilistic Risk Assessment (PRA)

Probabilistic risk assessment (PRA), also known as probabilistic safety assessment in certain contexts, is a systematic and comprehensive methodology used to evaluate and quantify the risks associated with complex systems, processes, and activities. It is widely applied in various industries, including nuclear power, aerospace, chemical manufacturing, and healthcare, to assess and manage risks. PRA provides a structured approach to understanding the likelihood and consequences of adverse events or failures within a system. Probabilistic risk assessment is a powerful tool for systematically understanding and managing risks in complex systems. It helps organizations make informed decisions to enhance safety, reliability, and overall performance while minimizing the likelihood of adverse events [3, 4]. The key components and concepts of PRA are as follows:

- Identification of Initiating Events: PRA starts by identifying and categorizing the initiating events or conditions that can lead to undesired outcomes or failures in a system. These events can be equipment failures, human errors, natural disasters, or other factors.
- Event Tree Analysis: Event tree analysis is a graphical representation that explores the possible sequences of events following an initiating event. It helps in understanding the various potential pathways that lead to different outcomes.

- Fault Tree Analysis: Fault tree analysis is another graphical tool used in PRA. It analyzes the various combinations of basic events and conditions that can lead to a specific undesired event or system failure. It is often used to assess the reliability and safety of complex systems.
- Quantification of Basic Events: As mentioned earlier, PRA involves quantifying the probabilities and consequences of basic events. This includes estimating the likelihood of individual events and the severity of their consequences.
- System Modeling: PRA requires the development of mathematical models that represent the system being analyzed. These models take into account system components, interactions, dependencies, and the dynamic nature of events.
- Data Collection and Analysis: Gathering data, including historical data and expert opinions, is essential for quantifying probabilities and consequences. Statistical methods and engineering judgment are often used to analyze and interpret the data.
- Risk Assessment: PRA calculates the risk associated with different scenarios and outcomes. The risk is typically expressed as a numerical value that combines the probability of occurrence and the severity of consequences.
- Sensitivity and Uncertainty Analysis: PRA accounts for uncertainties in the data and models. Sensitivity analysis helps identify which factors have the most significant impact on risk results, whereas uncertainty analysis quantifies the overall uncertainty in the risk estimates.
- Risk Mitigation and Decision Support: PRA results provide valuable information to decision-makers for risk-informed decision-making. It helps identify areas where risk mitigation measures are needed and supports resource allocation for safety improvements.
- Documentation and Reporting: Comprehensive documentation of the PRA process, including assumptions, methodologies, and results, is crucial for transparency and regulatory compliance. Reports are often submitted to regulatory authorities or used for internal decision-making.
- Continuous Improvement: PRA is an ongoing process, and as new data and insights become available, it is essential to update and refine the risk assessment to ensure that risk management strategies remain effective.

11.4.1 Importance of Probabilistic Risk Assessment

Probabilistic risk assessment plays a vital role in improving safety, optimizing resource allocation, ensuring regulatory compliance, and fostering a culture of risk awareness and mitigation in various industries. Its importance extends to both safety-critical industries and those seeking to proactively manage risks for improved operational and financial outcomes. PRA in certain contexts is of significant importance in various industries and applications due to the following reasons:

- Enhanced Safety: PRA is primarily used to assess and manage risks associated with complex systems and processes. By identifying potential hazards, quantifying risks, and prioritizing safety measures, PRA helps enhance safety and reduce the likelihood of accidents, incidents, or failures.
- Informed Decision-Making: PRA provides decision-makers with valuable information to make informed choices about safety measures, resource allocation, and risk acceptance. This enables organizations to prioritize actions based on the highest risks, ensuring that limited resources are used effectively.
- Resource Allocation: PRA allows organizations to allocate resources strategically. By identifying and quantifying risks, organizations can focus their efforts on mitigating or managing the most significant risks, optimizing resource allocation for safety improvements.
- Regulatory Compliance: In regulated industries such as nuclear power, aviation, and chemical manufacturing, PRA is often required by regulatory authorities to demonstrate compliance with safety standards and regulations. Conducting PRA helps organizations meet these regulatory requirements.
- Continuous Improvement: PRA is an iterative process that encourages continuous improvement in safety and risk management. Organizations can update and refine their risk assessments as new data becomes available or as systems evolve, ensuring that risk management strategies remain effective over time.
- Risk Communication: PRA provides a structured and transparent way to communicate risks to various stakeholders, including regulatory agencies, the public, and internal

teams. This enhances transparency and accountability in risk management.

- Complex System Understanding: In complex systems, it can be challenging to identify all potential risks and their interactions. PRA helps organizations gain a deeper understanding of the system's vulnerabilities, leading to more effective risk mitigation.
- Cost Reduction: Identifying and addressing risks early in the design or operational phase of a system can be cost-effective. PRA helps organizations avoid costly incidents, downtime, and damage repairs by proactively managing risks.
- Safety Culture: Implementing PRA fosters a safety culture within an organization, emphasizing the importance of safety at all levels. It encourages employees to be vigilant about identifying and reporting potential risks and hazards.
- Emergency Preparedness: PRA supports emergency preparedness by identifying scenarios that may lead to critical events. This helps organizations develop and implement effective emergency response plans to mitigate the consequences of such events.
- Public Confidence: For industries with public visibility, such as nuclear power and aerospace, conducting PRA and transparently sharing risk information can help maintain public confidence and trust in the safety of these technologies and systems.
- Reduced Liabilities: By identifying and addressing potential risks in advance, organizations can reduce their legal liabilities and the financial consequences of accidents or failures.

11.4.2 Advantages and Disadvantages of Probabilistic Risk Assessment

Organizations often balance the advantages and disadvantages of PRA by using a combination of quantitative and qualitative risk assessment methods. This hybrid approach allows for a more holistic understanding of risks, considering both data-driven quantification and expert judgment when necessary. The choice between quantitative and qualitative methods depends on the specific context, data availability, and goals of the risk assessment. However, like any approach, it important to consider the set of advantages and disadvantages, which are as follows

Advantages:

- Comprehensive Risk Assessment: PRA provides a systematic and comprehensive approach to risk assessment, considering a wide range of potential scenarios and outcomes. It helps identify and prioritize risks effectively.
- Quantitative Analysis: PRA quantifies risks, allowing organizations to assign numerical values to the likelihood and consequences of events. This enables more precise risk comparisons and decision-making.
- Risk Prioritization: PRA helps prioritize risks by assessing their potential impact. This allows organizations to focus resources and efforts on addressing the most significant and critical risks first.
- Transparency: PRA encourages transparency in risk assessment by documenting assumptions, data sources, and methodologies used. This transparency is essential for communicating risk information to stakeholders and regulatory authorities.
- Risk-Informed Decision-Making: PRA results provide valuable information for decision-makers, enabling them to make informed choices about safety measures, resource allocation, and risk acceptance.
- Sensitivity Analysis: PRA allows for sensitivity analysis, which helps identify which factors have the most significant influence on risk results. This insight is valuable for refining risk management strategies.
- Continuous Improvement: PRA is an iterative process. It supports ongoing monitoring and reassessment of risks as new data becomes available or as systems evolve, facilitating continuous improvement in risk management.

Disadvantages:

- Resource-Intensive: PRA can be resource-intensive in terms of time, expertise, and costs. Collecting data, building models, and conducting analyses can require significant resources.
- Data Requirements: PRA relies on data, and, in some cases, relevant data may be limited or unavailable, leading to uncertainties in risk assessments.

- Complexity: PRA is complex and may be challenging for non-experts to understand. It often requires specialized knowledge in risk analysis and modeling.
- Subjectivity: Despite its quantitative nature, PRA can still be influenced by subjective judgments, especially when expert opinions are used to fill data gaps.
- Model Assumptions: PRA models are based on assumptions, and small changes in these assumptions can lead to significantly different risk estimates. It is crucial to carefully consider and validate assumptions.
- Regulatory Burden: In regulated industries, conducting PRA and complying with specific regulatory requirements can be burdensome and time-consuming.
- Overemphasis on Quantification: While quantification is valuable, there is a risk of overemphasizing the quantitative aspect of risk assessment at the expense of qualitative insights and expert judgment.
- Complexity and Uncertainty Communication: Communicating PRA results to non-technical stakeholders can be challenging, especially when dealing with complex models and uncertainties.

11.5 Quantification of Basic Events

Quantification of basic events is a fundamental step in PRA and is often used in various industries, including nuclear power, chemical manufacturing, and aerospace, to assess and manage risks associated with complex systems and processes. Basic events are the lowest-level events or conditions that contribute to a specific undesired outcome or system failure. Quantifying these basic events involves assigning probabilities and consequences to them. Quantification of basic events helps organizations make informed decisions about risk management, resource allocation, and safety improvements. It provides a structured and systematic approach to understanding and addressing risks within complex systems and processes [5].

11.5.1 Importance of Quantification of Basic Events

Quantification of basic events plays a central role in various industries and fields where safety and risk management are paramount enabling organizations to systematically assess, prioritize, and address risks while enhancing safety, reliability, and overall performance. It is a crucial tool for achieving safety goals and regulatory compliance in various industries. Some key reasons why quantification of basic events is important are as follows:

- Risk Assessment: Quantification allows for a systematic and quantitative evaluation of the risks associated with complex systems or processes. It provides a clear understanding of the likelihood and consequences of various events, enabling organizations to prioritize and address high-risk areas.
- Informed Decision-Making: Quantified risk assessments provide decision-makers with valuable information for making informed choices. It helps them allocate resources, set priorities, and implement risk mitigation measures where they are needed most.
- Resource Allocation: By quantifying the risks associated with different basic events, organizations can allocate resources efficiently. They can focus their efforts and investments on mitigating or managing the most significant risks, thereby optimizing resource utilization.
- Regulatory Compliance: In regulated industries, such as nuclear power, aviation, and healthcare, quantified risk assessments are often required to demonstrate compliance with safety standards and regulations. Accurate quantification helps organizations meet these regulatory requirements.
- Continuous Improvement: Quantified risk assessments are not static but evolve over time as new data and insights become available. This allows organizations to continuously improve safety and risk management strategies based on updated risk assessments.
- Prioritization: Quantification helps in prioritizing basic events based on their potential impact and likelihood. This prioritization ensures that attention is directed toward addressing the most critical risks first.
- Resource Efficiency: It helps organizations avoid over-investing in low-probability, low-consequence events while ensuring that they are adequately prepared for high-probability, high-consequence events. This efficiency is particularly important in resource-constrained environments.
- Communication: Quantified risk assessments provide a common language for discussing risks within an organization and with external stakeholders. It facilitates clear and

effective communication about the relative importance of different risks.

- Safety Improvement: Identifying and quantifying basic events allow organizations to design and implement safety improvements proactively. It is often more cost-effective to address potential risks during the design or planning phase than after an incident has occurred.
- Accountability: Quantified risk assessments help establish accountability within organizations. They make it clear who is responsible for managing and mitigating specific risks and ensure that actions are taken to reduce risk to acceptable levels.
- Transparency: Quantification brings transparency to the risk assessment process. It allows stakeholders, including employees, regulators, and the public, to understand the basis for risk management decisions and to have confidence in the safety of the systems or processes involved.

11.5.2 Procedures Involved in Quantifying Basic Events

Quantification of basic events is typically conducted to assess and measure the likelihood and consequences of individual events or conditions that may contribute to system failures or undesired outcomes. Quantification of basic events provides a foundation for understanding and managing risks in complex systems or processes. It supports informed decision-making and helps organizations prioritize risk mitigation efforts to enhance safety and reliability. General steps and procedures involved in quantifying basic events are as follows:

(i) Event Identification: Begin by identifying and documenting the basic events, also known as initiating events, that can trigger or contribute to undesired outcomes. These events should be specific, well-defined, and relevant to the system or process being analyzed.

(ii) Categorization and Classification: Categorize the identified events based on their characteristics, causes, and effects. Events can be classified into various categories, such as equipment failures, human errors, external hazards, or system malfunctions.

(iii) Data Collection: Gather relevant data and information about each basic event. This may involve historical data, engineering specifications, expert opinions, and any available documentation related to the events.

(iv) Quantitative and Qualitative Assessment: Depending on the available data and the complexity of the analysis, you can perform either quantitative or qualitative assessments of basic events.

- Quantitative Assessment: In cases where sufficient data is available, use mathematical models and statistical methods to quantify the probability of each basic event occurring. This may involve the use of failure data, reliability analysis, and event tree or fault tree analysis.
- Qualitative Assessment: In situations where quantitative data is limited, rely on expert judgment and qualitative assessments to estimate the likelihood of each event occurring. This often involves expert workshops or structured interviews to gather input from subject matter experts.

(v) Probabilistic Analysis: If quantitative data is available, then use probability distributions and statistical techniques to estimate the probabilities of basic events. Common methods include probability density functions, cumulative distribution functions, and Bayesian analysis.

(vi) Consequence Analysis: Determine the consequences associated with each basic event if it were to occur. Consequences can vary widely and may include harm to people, environmental damage, equipment damage, financial losses, and operational disruptions.

(vii) Consequence Quantification: Assign numerical values or ranges to the consequences associated with each basic event. These values may represent severity, cost, or other relevant metrics.

(viii) Risk Calculation: Calculate the risk associated with each basic event by multiplying the estimated probability of occurrence by the associated consequence. This results in a risk value, often expressed as a numerical score or ranking.

(ix) Risk Ranking and Prioritization: Rank the basic events based on their calculated risk values. This prioritization helps focus attention and resources on mitigating or managing the most significant risks.

(x) Sensitivity Analysis: Conduct sensitivity analysis to assess how variations in probability and consequence estimates impact the overall risk assessment. This helps identify which factors have the most significant influence on risk results.

(xi) Documentation and Reporting: Maintain clear and detailed documentation of the quantification process, including assumptions, methodologies, data sources, and calculations. Communicate the results to relevant stakeholders and decision-makers.

(xii) Continuous Review and Improvement: Quantification of basic events is an iterative process. Continuously monitor and update the assessments as new data or insights become available, and periodically review and revise risk management strategies accordingly.

11.5.3 Advantages and Disadvantages of Quantification of Basic Events

Quantification of basic events offers several advantages and disadvantages, depending on the context and the specific application. Organizations often use a combination of quantitative and qualitative risk assessment methods to address these advantages and disadvantages. This hybrid approach allows for a more holistic understanding of risks, taking into account both data-driven quantification and expert judgment when necessary. The choice between quantitative and qualitative methods depends on the specific context, the availability of data, and the goals of the risk assessment. Some of the key advantages and disadvantages are as follows:

Advantages:

- Systematic Approach: Quantification of basic events provides a systematic and structured approach to assessing and managing risks. It helps ensure that all relevant risks are considered and evaluated.
- Objective Decision-Making: It allows for objective decision-making by quantifying risks in terms of probabilities and consequences, reducing reliance on subjective judgments.
- Prioritization: By assigning numerical values to risks, organizations can prioritize their efforts and resources

on mitigating or managing the most significant risks first, improving resource allocation.

- Transparency: The process of quantification is transparent and well-documented, making it easier to communicate risk information to stakeholders, including regulatory authorities and the public.
- Risk-Informed Decision-Making: Quantified risk assessments provide valuable information to decision-makers, enabling them to make informed choices about safety measures, resource allocation, and risk acceptance.
- Continuous Improvement: The quantitative approach supports ongoing monitoring and reassessment of risks as new data becomes available, facilitating continuous improvement in risk management.

Disadvantages:

- Data Limitations: Accurate quantification of risks relies on quality data, and, in some cases, relevant data may be limited or unavailable. This can introduce uncertainty into the assessment.
- Complexity: The process of quantifying basic events can be complex, particularly for large and intricate systems. It requires specialized knowledge and resources.
- Resource-Intensive: Quantitative risk assessments can be resource-intensive, both in terms of time and costs. Gathering data, conducting analyses, and maintaining models can be expensive.
- Sensitivity to Assumptions: The results of quantitative risk assessments can be sensitive to the assumptions made during the modeling and analysis process. Small changes in assumptions can lead to significantly different risk estimates.
- Incompleteness: Quantification of basic events may not capture all possible scenarios or combinations of events. There can be limitations in how well the assessment accounts for rare or unique events.
- Communication Challenges: Communicating the results of quantitative risk assessments to non-technical stakeholders can be challenging. The numerical nature of the results may be difficult for some individuals to interpret.

- Regulatory Burden: In regulated industries, there may be a significant regulatory burden associated with conducting and documenting quantitative risk assessments, which can increase administrative overhead.
- Overemphasis on Quantification: While quantification is valuable, there is a risk of overemphasizing the quantitative aspect of risk assessment at the expense of qualitative insights and expert judgment.

References

1. Dhalmahapatra, K., Verma, A., Maiti, J., An integrated TRIZ coupled safety function deployment and capital budgeting methodology for occupational safety improvement: A case of manufacturing industry. *Process Saf. Environ. Prot.*, 165, 31–45, 2022.
2. Fargnoli, M., Lombardi, M., Haber, N., Guadagno, F., Hazard function deployment: A QFD-based tool for the assessment of working tasks–A practical study in the construction industry. *Int. J. Occup. Saf. Ergon.*, 26, 2, 348–369, 2018.
3. Henley, E.J. and Kumamoto, H., *Probabilistic risk assessment and management for engineers and scientists*, IEEE Press, United States, 1996.
4. Rasmussen, N.C., The application of probabilistic risk assessment techniques to energy technologies, in: *Readings in Risk*, pp. 195–206, 2013.
5. Vaurio, J.K., Uncertainties and quantification of common cause failure rates and probabilities for system analyses. *Reliab. Eng. Syst. Saf.*, 90, 2-3, 186–195, 2005.

12

Human Errors: Classification, Causes, and Identification

Abstract
This chapter describes human errors and the factors that lead to them. It also discusses the different classifications of human error with examples. Different methods to identify those errors have also been added. In the latter part, a detailed discussion was provided about important steps that can be beneficial in preventing such human errors.

***Keywords*:** Human errors, mistakes, identification, prevention

12.1 Background

In every aspect of our lives, from personal endeavors to professional pursuits, mistakes are an inevitable part of being human. However, when it comes to the workplace, the cost of human errors can be significant. These errors can lead to accidents, incidents, and potentially harmful consequences for individuals and businesses, whether due to slips, lapses, or mistakes. That is why understanding the different types of human errors and implementing preventive measures is crucial to maintaining a safe and productive work environment. By identifying the root causes of these errors, such as fatigue, stress, distractions, or inadequate training, we can develop proactive measures to mitigate their occurrence. By fostering a culture of safety and providing the necessary tools and knowledge, we can empower individuals to minimize the likelihood of errors and promote a safer workplace for all.

Abhishek Barua, Kanchan Kumari, Sumit Saha, Siddharth Jeet and Swastik Pradhan. Institutional and Industrial Safety Engineering Practices, (287–302) © 2025 Scrivener Publishing LLC

12.2 Introduction

When all other causes have been excluded, human error is frequently stated as a cause of accidents. This is not to say that scientific concepts cannot be used to analyze human mistakes. In fact, there is a lot of interest, nowadays, in studying human mistakes. Human error cannot be completely eradicated, but, by identifying common errors, the majority of them may be avoided.

Human errors, according to conventional wisdom, are a source of failure and accidents. Human errors, according to a new philosophical viewpoint, are a symptom of failure that reveals underlying issues in a system. Examining human errors gives us knowledge that allows us to go beyond the simplistic label. Human errors are an after-the-fact attribution that is systematically tied to people, tools, tasks, and the operational environment. Although there is no universal definition of human errors, the general consensus has shifted away from assigning blame to individuals and toward a much broader contextual approach. The unintentional character of such errors distinguishes them from purposeful acts (known as "violations"), in which a person willingly and consciously takes an improper course of action [1].

12.3 Causes of Human Error

Human errors can be influenced by a wide range of factors, and understanding these factors is essential for preventing or mitigating errors in various contexts. Addressing these factors often involves a combination of strategies, including improved training programs, managing workload and schedules, fostering a culture of safety and open communication, providing clear and detailed procedures, and implementing systems that reduce the impact of human errors, such as automation or error-proofing mechanisms [1]. Some common causes of human errors include the following:

- Lack of Training: Insufficient training or inadequate understanding of procedures and tasks can lead to errors. Proper training ensures that individuals have the necessary knowledge and skills to perform their jobs effectively and safely.
- Lack of Experience: Inexperienced individuals are more likely to make mistakes, as they may not have encountered or learned how to handle specific situations.

- Inadequate Supervision: Lack of supervision, or ineffective supervision, can result in employees not following procedures or taking shortcuts, leading to errors.
- Fatigue: Fatigue, whether due to lack of sleep, long working hours, or mental exhaustion, impairs concentration, decision-making, and reaction times, increasing the likelihood of errors.
- Stress: High levels of stress, whether from work pressure, personal issues, or challenging situations, can impair cognitive function, attention, and memory, making individuals more prone to mistakes.
- Lack of Standard Procedures: Absence of clear, well-defined procedures or guidelines can lead to confusion and errors, especially in situations that require specific steps to be followed.
- Distractions: External distractions (e.g., noisy environments and interruptions) or internal distractions (e.g., wandering thoughts) divert attention away from tasks, leading to errors.
- Lack of Communication: Poor communication, misunderstandings, or unclear instructions can result in misinterpretation and subsequent errors in tasks and decision-making.
- Complacency: Overconfidence or complacency arising from routine tasks can lead to a decrease in vigilance and attention, making individuals more susceptible to errors. Excessive confidence in one's abilities can lead to complacency and a reduced focus on tasks, increasing the risk of errors.
- Time Pressure: Working under tight deadlines or time constraints can lead to rushing, causing individuals to skip steps, overlook details, or make impulsive decisions, resulting in errors.
- Complexity of Tasks: Complex tasks with multiple steps and decision points increase the likelihood of errors, especially if the procedures are not well-defined or are too intricate.
- Inadequate Feedback: Lack of feedback on performance can prevent individuals from recognizing and correcting their errors, leading to the repetition of mistakes.
- Organizational Culture: A culture that does not prioritize safety, open communication, or learning from mistakes can contribute to an environment where errors are more likely to occur and less likely to be reported or corrected.

- Inadequate Tools and Equipment: Errors can also result from poorly designed tools, interfaces, or equipment that are not user-friendly or intuitive, leading to incorrect operation. Outdated or malfunctioning equipment, tools, or technology can create unexpected challenges and errors in tasks that require their use.
- Cognitive Biases: Human cognition is prone to biases such as confirmation bias (favoring information that confirms previous ideas) and the availability heuristic (relying on easily accessible information). These biases might result in poor decision-making and mistakes.

12.4 Identification of Human Error

Human errors can be classified into different categories based on various criteria. Slips, lapses, and mistakes are terms used in the field of human factors to describe different types of errors that individuals can make. Understanding the distinctions between these types of errors can help in designing systems and procedures to prevent or mitigate them effectively [2]. Figure 12.1 represents categorization of human errors.

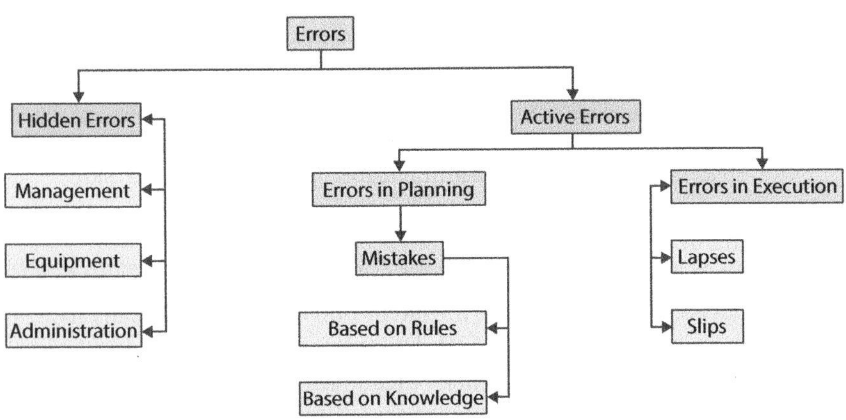

Figure 12.1 Categorization of human errors.

12.4.1 Basic Types of Human Error

12.4.1.1 Slips

Slips are errors that occur during the execution of a task, where the intended action is different from the action actually performed.

Characteristics:

- Automatic Tasks: Slips typically occur during tasks that are routine and automatic, requiring minimal conscious attention.
- Inattention or Distraction: Slips often happen when a person's attention is diverted or when they are not fully focused on the task at hand.

Example: Pouring juice into a cereal bowl instead of a glass because both are located next to each other during breakfast.

12.4.1.2 Lapses

Lapses are errors of omission, where an individual forgets to perform a necessary action or fails to complete a step in a task.

Characteristics:

- Memory Failure: Lapses occur due to a failure in memory, leading to the omission of a step or action.
- Routine Tasks: Lapses can occur in tasks that are part of a person's routine but are missed due to memory lapses.

Example: Forgetting to turn off the stove after cooking because of a momentary lapse in memory.

12.4.1.3 Mistakes

Definition: Mistakes are errors that occur when there is a failure in the decision-making or problem-solving process, leading to incorrect actions or choices.

Characteristics:

- Knowledge or Rule-Based: Mistakes can be rule-based (applying the wrong rule) or knowledge-based (lack of knowledge or incorrect understanding).
- Conscious Decision: Mistakes involve a conscious decision or choice, but the decision is flawed due to incorrect information or reasoning.

Example: Taking the wrong turn while driving because of misinterpreting the directions or lacking knowledge of the correct route.

Addressing slips, lapses, and mistakes often involves different strategies. Slips can be reduced through improved design, such as making controls more distinct to avoid confusion. Lapses can be mitigated by implementing reminders, checklists, or alarms to prompt individuals to perform necessary actions. Mistakes can be minimized through education, training, and providing clear and detailed information to enhance decision-making processes. Creating environments that support attention, memory, and accurate decision-making is essential in reducing the occurrence of these errors.

12.4.2 Active Errors vs. Latent Errors

Active errors and latent errors are terms used in the field of human factors and safety engineering to describe different aspects of errors that can occur in complex systems. These concepts are often associated with the investigation and analysis of accidents and incidents. Active errors are the immediate mistakes made by individuals, whereas latent errors are underlying issues within the system that can contribute to or amplify active errors. Understanding both types of errors is crucial for improving system safety and preventing accidents. Effective safety management involves identifying and addressing both active and latent errors to create a safer and more reliable system.

12.4.2.1 Active Errors

Active errors, also known as immediate or direct errors, are errors that occur at the frontline, where individuals interact with the system or process. These errors are typically made by people who are directly involved in the task and can have an immediate impact on the system's performance or safety. Active errors are often visible and can be observed in real-time.

Example: A pilot making an incorrect maneuver during a flight, a nurse administering the wrong medication, or a driver running a red light due to distraction.

12.4.2.2 Latent Errors

Latent errors, also known as systemic errors or organizational errors, are inherent within the system's design, policies, or organizational culture. Unlike active errors, latent errors are not immediately obvious and may remain dormant within the system for a long time. These errors are usually the result of decisions made by system designers, management, or policy-makers. Latent errors can contribute to active errors or create conditions where active errors are more likely to occur.
Example: Flawed system design, inadequate training programs, unclear communication protocols, or budget cuts leading to staffing shortages. These factors might not cause immediate issues but can create an environment where active errors are more likely to happen.

12.4.2.3 Relationship Between Active and Latent Errors

- Error Pathway: Latent errors can create the conditions or set the stage for active errors to occur. For instance, a poorly designed user interface (latent error) might confuse a user, leading to a data entry mistake (active error).
- Amplification: Latent errors, if not identified and rectified, can lead to repeated active errors, exacerbating the potential consequences. For example, inadequate training (latent error) might result in multiple instances of the same mistake by various employees (active errors).
- Intervention Points: Addressing latent errors often involves systemic changes, such as redesigning processes, improving training, or changing organizational policies. Addressing active errors might involve immediate corrective actions, such as retraining individuals or implementing additional safety checks.

12.4.3 Errors Based on Decision, Skill, or Perception

Human errors can be categorized into different types based on decision-making, skill execution, or perception. Addressing these types of errors often involves targeted interventions.

12.4.3.1 Decision-Based Errors

Decision-based errors occur when individuals make faulty judgments or decisions. These errors are often influenced by cognitive biases or incomplete information. Decision-based errors can be mitigated through training programs that focus on recognizing and overcoming cognitive biases.

Examples:

- Confirmation Bias: Focusing on information that confirms preexisting beliefs while ignoring contradictory evidence
- Overconfidence Bias: Being excessively confident in one's abilities or judgments, leading to risky decisions.
- Anchoring Bias: Relying too heavily on the first piece of information encountered when making decisions.
- Hindsight Bias: Perceiving events as having been more predictable after they have occurred.

12.4.3.2 Skill-Based Errors

Skill-based errors happen during routine tasks that a person has performed many times. These errors occur due to lapses in attention or motor skills. Skill-based errors can be reduced through improved training, clear procedures, and environmental design that minimizes distractions.

Examples:

- Slips: Mistakes made during automatic tasks due to a failure to execute the correct action (e.g., pressing the wrong button on a familiar device).
- Lapses: Forgetting to perform an action, even though it is a routine part of the task (e.g., forgetting to turn off an appliance).

12.4.3.3 Perception-Based Errors

Perception-based errors occur when individuals misinterpret sensory information, leading to incorrect judgments or actions. Perception-based errors can be minimized by providing clear and unambiguous information, enhancing communication, and addressing potential sources of sensory confusion.

Examples:

- Illusions: Misinterpreting a sensory stimulus (e.g., optical illusions) that leads to an incorrect perception of reality.
- Ambiguity: Misinterpreting vague or unclear information, leading to misjudgments or mistakes.

12.4.3.4 Knowledge-Based Errors

Knowledge-based errors occur when individuals lack the necessary information or expertise to make correct decisions or perform a task accurately. Knowledge-based errors can be reduced through ongoing education, access to accurate information, and mentorship or guidance from experts in the field.

Examples:

- Misapplication of Rules: Applying the wrong rule or guideline in a specific situation due to a lack of knowledge or understanding.
- Misinterpretation of Data: Incorrectly interpreting data or information due to a lack of expertise in the subject matter.

12.4.4 Errors Based on Psychological Factors

Human errors influenced by psychological factors can be categorized in various ways based on the underlying psychological processes. Addressing errors influenced by psychological factors involves awareness, training, and creating environments that minimize the impact of cognitive biases, distractions, emotional states, social pressures, and motivational factors. Training programs, feedback mechanisms, and promoting a culture of mindfulness can help individuals recognize and mitigate these psychological influences, reducing the likelihood of errors.

12.4.4.1 Cognitive Biases

Errors caused by cognitive biases, which are continuous patterns of divergence from norm or rationality in judgement, frequently leading to perceptive distortion, incorrect judgement, illogical analysis, or what is referred to as irrationality.

Examples:

- Confirmation Bias: The tendency to interpret or recall facts in a way that supports one's assumptions.
- Accessibility Heuristic: Calculating the likelihood of an occurrence based on its memory accessibility.
- Anchoring Bias: When making judgements, rely too strongly on the initial piece of information received.

12.4.4.2 Attentional Errors

Errors arising due to limited attentional resources, distractions, or information overload, leading to overlooking critical details.

Examples:

- Inattentional Blindness: Failing to perceive an unexpected stimulus in plain sight due to attention being focused elsewhere.
- Distraction Errors: Making mistakes because attention is diverted away from the task at hand.

12.4.4.3 Memory Errors

Errors occurring due to failures in encoding, storage, retrieval, or distortion of information in memory.

Examples:

- Misattribution: Assigning a memory to the wrong source, resulting in false memories.
- Transience: Forgetting information over time, leading to inaccuracies in decision-making.

12.4.4.4 Emotional Errors

Errors influenced by emotional states, such as stress, fear, or anger, which can impair cognitive functions and lead to poor decision-making.

Examples:

- Emotional Impairment: Making impulsive decisions under intense emotional states, leading to regrettable outcomes.

- Stress-Related Errors: Errors caused by heightened stress levels, affecting attention, memory, and judgment.

12.4.4.5 Social Influence Errors

Errors resulting from conformity, obedience, or groupthink, where individuals make decisions influenced by social pressure or authority figures.

Examples:

- Conformity Errors: Going along with the majority opinion even if it contradicts personal beliefs or knowledge.
- Authority Errors: Blindly following the instructions of an authority figure, even if it leads to incorrect decisions.

12.4.4.6 Motivational Errors

Errors influenced by motivational factors, such as incentives, rewards, or punishments, which can lead to biased decision-making.

Examples:

- Reward-Induced Bias: Making decisions based on the desire to receive rewards, even if they are suboptimal choices.
- Fear-Driven Errors: Making hasty or irrational decisions due to the fear of negative consequences.

12.4.5 Errors Based on Violations

Violations, in the context of human errors, refer to intentional deviations from established rules, procedures, or norms. Unlike other types of errors, violations are conscious actions where individuals deliberately choose to violate the rules or guidelines. Strategies to prevent violations include clear communication of rules, providing adequate training and resources, fostering a positive safety culture, addressing peer pressure and social norms, and creating a system that discourages and penalizes intentional violations while encouraging open communication about potential issues.

12.4.5.1 Routine Violations

Deviations from established rules or procedures that occur as part of a person's regular behavior. These violations are predictable and often become an accepted part of the work culture.

Example: Ignoring safety protocols because employees believe they are unnecessary for a routine task.

12.4.5.2 *Situational Violations*

Deviations from rules or procedures that occur in response to specific circumstances or situations. These violations are often driven by a perceived need to achieve a specific goal or address an immediate problem.
Example: Bypassing safety procedures to meet a tight deadline during an emergency situation.

12.4.5.3 *Exceptional Violations*

Rare, extreme violations that occur under highly exceptional circumstances. These violations are not typical of an individual's behavior and often result from extreme pressure, stress, or unusual conditions.
Example: Completely disregarding safety protocols due to overwhelming panic during a crisis.

12.4.5.4 *Optimizing Violations*

Violations that occur when individuals deliberately deviate from rules or procedures to optimize their performance or achieve a more efficient outcome. These violations are driven by a belief that the deviation will lead to better results.
Example: Using a shortcut in a process to save time, even though it violates established procedures.

12.4.5.5 *Social Violations*

Violations that occur due to social or peer pressure. Individuals may violate rules or norms because they perceive that others in their social group are doing the same, leading to conformity.
Example: Violating workplace policies because colleagues engage in similar behavior without facing consequences.

12.4.5.6 *Gaming the System*

Manipulating or exploiting loopholes in rules or procedures to achieve personal gain or advantage, often without technically violating the written rules but contradicting the intended spirit of the rules.

Example: Exploiting ambiguities in policies to gain unfair advantages in a competitive setting.

12.4.6 Methods to Identify Human Errors

Identifying human errors is a crucial step in improving processes, systems, and overall safety in various contexts, including workplaces, healthcare, transportation, and everyday activities. Combining multiple methods and tools often provides a more comprehensive understanding of human errors and their causes. By addressing the root causes of human errors, organizations can implement preventive measures, such as improved training, process redesign, automation, or changes in organizational culture, to minimize the occurrence of errors and enhance overall performance and safety. There are several methods and tools available to identify human errors:

1. Root Cause Analysis (RCA): RCA is a systematic process for identifying the underlying causes of problems or events. By investigating an incident, you can identify the human errors that contributed to it. Techniques like the "5 Whys" method involve asking "why" repeatedly to drill down to the fundamental cause, which might be a human error.
2. Failure Mode and Effects Analysis (FMEA): FMEA is a structured approach to discovering potential failures in a process, product, or system and then determining their impact. Human errors are often included as potential failure modes. By assessing the severity, occurrence, and detection of human errors, organizations can prioritize and address them.
3. Human Factors Analysis and Classification System (HFACS): HFACS is a framework used in aviation and other high-risk industries to identify human errors and their contributing factors. It categorizes errors into four levels: unsafe acts, preconditions for unsafe acts, unsafe supervision, and organizational influences. This method provides a comprehensive view of human errors and their causes.
4. Observational Studies: Observing people while they perform tasks can reveal common errors and areas for improvement. These observations can be done in real work environments or simulated settings. By analyzing the observations, patterns of human errors can be identified.

5. Surveys and Interviews: Surveys and interviews can be conducted to gather information from employees about their experiences and perceptions of errors in the workplace. This qualitative data can provide insights into the types of errors that occur and the factors contributing to them.
6. Checklists and Incident Reports: Analyzing incident reports and checklists can help identify recurring errors. Incident reports often contain valuable information about the circumstances, actions, and decisions leading to accidents or mistakes.
7. Task Analysis: Breaking down tasks into smaller steps and analyzing each step can reveal potential points of error. By understanding the cognitive and physical demands of tasks, organizations can design processes that reduce the likelihood of human errors.
8. Simulation and Modeling: Using computer simulations and modeling human behavior in virtual environments can help identify potential errors in complex systems. By simulating various scenarios, organizations can anticipate and mitigate human errors before they occur.

12.5 Prevention of Human Error

Preventing human errors involves a multi-faceted approach that combines strategies related to training, system design, communication, and organizational culture. Organizations can create a safer and more reliable environment, reducing the occurrence of human errors and enhancing overall performance and safety. There are several key methods to prevent human errors in various contexts [3].

12.5.1 Training and Education

- Clear Procedures: Provide comprehensive and clear instructions for tasks, processes, and protocols to ensure employees know the correct procedures to follow.
- Regular Training: Conduct regular training sessions to refresh skills, reinforce knowledge, and introduce new best practices.

- Simulation and Drills: Use simulations and practice drills to allow individuals to experience different scenarios and learn how to respond effectively.

12.5.2 System Design and Automation

- Simplification: Simplify complex processes and tasks to reduce the likelihood of errors. Minimize the number of steps and decision points.
- Automation: Automate repetitive or critical tasks to reduce the reliance on human performance, especially for tasks where consistency and precision are crucial.

12.5.3 Effective Communication

- Clear Communication: Ensure that instructions, procedures, and important information are communicated clearly, using simple language and visual aids when necessary.
- Open Communication Culture: Foster an environment where employees feel comfortable reporting errors and near misses without fear of punishment, encouraging learning from mistakes.

12.5.4 Workplace Environment

- Ergonomic Design: Design workspaces and tools ergonomically to minimize physical strain and reduce the risk of errors caused by discomfort or fatigue.
- Minimize Distractions: Reduce distractions in the workplace to help employees stay focused on their tasks.

12.5.5 Checklists and Procedures

- Checklists: Implement checklists for critical tasks to ensure that all necessary steps are completed. Pilots and surgeons, for instance, often use checklists to reduce errors.
- Standard Operating Procedures: Develop and enforce standardized operating procedures to provide a consistent framework for tasks and decision-making.

12.5.6 Human Factor Training

- Understanding Human Factors: Provide training on human factors and how cognitive biases, stress, and fatigue can influence decision-making and performance.
- Error Reporting: Encourage reporting of errors and near misses, allowing organizations to analyze patterns and implement preventive measures.

12.5.7 Continuous Improvement

- Root Cause Analysis: Conduct thorough investigations using techniques like RCA to understand the underlying causes of errors and implement systemic changes.
- Learning Organization: Foster a culture of continuous learning, where feedback from errors is used to improve processes and prevent future mistakes.

12.5.8 Leadership and Management

- Leadership Support: Leadership should demonstrate a commitment to safety and error prevention, setting an example for the rest of the organization.
- Accountability: Hold individuals accountable for their actions while also recognizing that errors often indicate system flaws rather than individual shortcomings.

References

1. Rouse, W.B. and Rouse, S.H., Analysis and classification of human error. *IEEE Trans. Syst. Man Cybern.*, 4, 539–549, 1983.
2. Kirwan, B., Human error identification in human reliability assessment. Part 1: Overview of approaches. *Appl. Ergon.*, 23, 5, 299–318, 1992.
3. Hu, W., Carver, J.C., Anu, V., Walia, G.S., Bradshaw, G.L., Using human error information for error prevention. *Empirical Software Eng.*, 23, 3768–3800, 2018.

13

Accident: Causes, Identification, and Investigation

Abstract
Although our society's emphasis on risk management grows, big incidents resulting in multiple fatalities appear to be inescapable in some industries. Because the repercussions of such significant accidents are unacceptably severe, a comprehensive study of the incidents should be conducted in order to learn from what occurred and prevent future mishaps. This chapter discusses the philosophy of accidents; the types, causes, and reasons that lead to accidents; and why they should be investigated. It also discusses the different basic formats of an accident report, the importance of accident investigation, different methods for investigation, and the structure of an accident investigation report. It also discusses the facts that should be kept in mind while doing an accident investigation, who should conduct the investigation, and what should be done if the accident is caused by human error.

Keywords: Accident, investigation, fact check, identification, human errors

13.1 Background

Any industrial operation is fraught with danger. Even if a robust risk management plan has been implemented to account for anticipated concerns, accidents can occur. It is a good idea to thoroughly examine any accidents or close misses. Accident investigation is an organized process that involves a formal plan to investigate the accident, document the investigation, track progress, and analyze investigation data in order to discover and control for recurrent accidents. A successful investigative technique is intended to establish solutions to deal with the underlying causes of accidents and near-misses rather than assign blame. Accident investigation continues to evolve, driven by technological advancements, interdisciplinary collaboration, and a growing emphasis on understanding the complex interplay between human factors and system failures. The ultimate goal remains

Abhishek Barua, Kanchan Kumari, Sumit Saha, Siddharth Jeet and Swastik Pradhan. Institutional and Industrial Safety Engineering Practices, (303–334) © 2025 Scrivener Publishing LLC

the prevention of accidents and the enhancement of safety across various industries and sectors.

13.2 Introduction

Every day, numerous accidents occur all over the world. Most accidents are caused by people, equipment, supplies, or the environment failing to behave or react as planned. Investigations into accidents establish how and why these errors occur. A similar or maybe more severe catastrophe could be avoided by using the knowledge obtained during an inquiry. Conduct accident investigations with the goal of preventing future accidents in mind. Investigations are not used to assign blame. An accident is defined as any unexpected incident that causes personal harm or property damage. A minor personal injury is one that requires little or no care. It is considered significant if it results in a death or a permanent total, permanent partial, or temporary total (lost-time) handicap. Property damage might either be modest or severe. Investigate all accidents, no matter how severe the injuries or damage. Accidents are one of several occurrences that might jeopardize the execution of a mission. These are occurrences. For clarity, the methods outlined in the following sections only apply to incidents. They are nevertheless relevant to situations as well [1]. This talk exposes the reader to fundamental accident investigation protocols and accident analysis methodologies.

13.3 What is an Accident and Why Should it be Investigated?

An accident is an unforeseen and unplanned event or circumstance that results in damage, injury, or harm to people, property, or the environment. Accidents can happen in various contexts, such as in transportation (car accidents and plane crashes), workplaces (industrial accidents), public spaces (slips and falls), or at home (fires, poisoning). Accidents can be caused by a wide range of factors, including human error, mechanical failure, environmental conditions, or a combination of these elements.

Accident investigation is the methodical gathering and analysis of facts pertaining to an event that resulted in the loss of property, time, or health of persons, or even the loss of lives. The scientific and intellectual study of the events that transpired during an accident is known as accident investigation. An investigation is carried out to determine the root cause of an accident in order to offer suggestions or take remedial steps to prevent the occurrence of the same or a similar incident in the future. Accident investigations demonstrate the need for safety programmes in every organization [1]. Finding the fundamental causes of accidents helps identify measures to prevent the same accident from happening again. Investigating accidents is essential to understanding their causes, preventing similar incidents, improving safety measures, ensuring accountability, and promoting overall public and workplace safety. Accident investigation is crucial for several reasons:

1. Prevention: Understanding the causes of an accident helps identify underlying issues and allows for corrective measures to be implemented. By addressing the root causes, similar accidents can be prevented in the future.
2. Safety Improvement: Investigations often reveal safety deficiencies in systems, equipment, or procedures. By addressing these shortcomings, safety can be improved, reducing the likelihood of future accidents.
3. Legal and Insurance Purposes: In legal and insurance matters, a thorough investigation provides valuable information for determining liability and settling claims. Accurate information about the accident is essential for legal proceedings and insurance claims.
4. Accountability: Investigations can determine if any individual or organization was at fault. This information can be crucial in legal cases and also ensures accountability, promoting responsible behavior in various contexts.
5. Learning and Knowledge Sharing: Through investigations, lessons can be learned not only by the parties directly involved but also by industries and society at large. Sharing the findings of investigations can lead to improvements in regulations, guidelines, and best practices, enhancing overall safety standards.
6. Public Safety: For accidents involving public transportation, infrastructure, or services, investigations help ensure public safety. Identifying and addressing systemic issues safeguard the well-being of citizens.

7. Organizational Improvement: Within organizations, analyzing accidents helps identify areas where processes can be improved, training can be enhanced, and safety cultures can be developed. This can lead to overall organizational improvement and a safer working environment.

13.4 Classification of Accident

Industrial accidents can be classified into several categories based on various criteria. Understanding these classifications helps industries and regulatory authorities analyze trends, implement targeted safety measures, and develop appropriate policies to prevent similar accidents in the future. Figure 13.1 depicts the basic classification of accident.

Some common classifications of industrial accidents are as follows:

a) By Severity
   - Minor Accidents: Incidents causing minor injuries or damage with short-term effects.

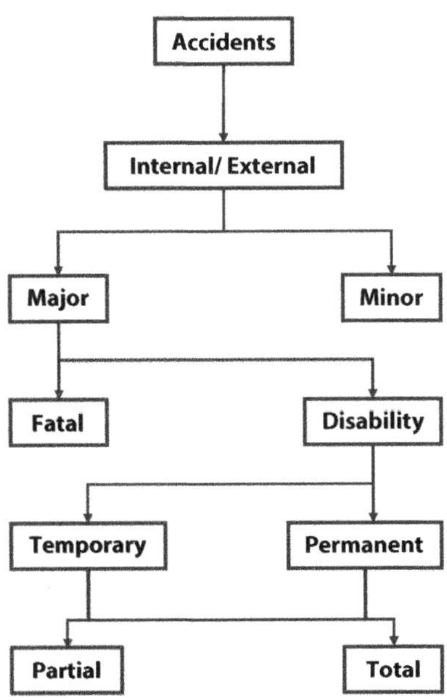

Figure 13.1 Basic classification of accidents.

- Major Accidents: Serious incidents leading to significant injuries, fatalities, or extensive damage with long-term effects.

b) By Type
   - Mechanical Accidents: Involving machinery or equipment malfunctions.
   - Chemical Accidents: Involving hazardous materials, spills, or leaks.
   - Physical Accidents: Involving slips, falls, or collisions.
   - Biological Accidents: Involving exposure to biological hazards.
   - Radiological Accidents: Involving exposure to radioactive materials.
   - Fire and Explosive Accidents: Involving fires, explosions, or combustible materials.

c) By Cause
   - Human Error Accidents: Caused by mistakes made by workers or management.
   - Equipment Failure Accidents: Caused by malfunctioning machinery or tools.
   - Environmental Accidents: Caused by natural disasters or environmental factors.
   - Structural Accidents: Involving building or structural failures.
   - Process Accidents: Resulting from failures in industrial processes or operations.

d) By Industry
   - Construction Accidents: Occurring in construction sites involving tools, heavy equipment, and heights.
   - Manufacturing Accidents: In industrial manufacturing settings involving machinery and production processes.
   - Chemical Industry Accidents: Involving the production, storage, or transportation of chemicals.
   - Mining Accidents: In mining operations, including collapses, gas leaks, or equipment failures.
   - Oil and Gas Industry Accidents: Involving drilling, refining, or transportation processes.

e) By Outcome
   - Fatal Accidents: Resulting in one or more fatalities.
   - Non-Fatal Accidents: Causing injuries but no fatalities.

- Property Damage Accidents: Involving damage to buildings, equipment, or other property.

f) By Frequency
- Common Accidents: Frequently occurring incidents in a specific industry or workplace.
- Rare Accidents: Uncommon incidents that happen infrequently but might have severe consequences.

g) By Preventability
- Preventable Accidents: Those that could have been avoided with proper precautions and safety measures.
- Non-Preventable Accidents: Resulting from unpredictable events or circumstances.

13.5 Different Types of Industrial Accidents

Industrial accidents are among the most tragic since they frequently result in personal injuries. This sort of event, as well as workplace accidents, can cause individuals to miss work or lose their jobs, depending on their injuries. Workers' compensation claims and lawsuits are frequently based on industrial accidents. Industrial accidents often occur when an employer fails to fulfil their obligation to safeguard their employees. This may be from failing to provide a safe environment, encouraging shortcuts in work, not training workers properly, or failing to supervise workers. To avoid industrial accidents, it is important to understand the different common types of industrial accidents. Industrial accidents can vary widely depending on the nature of the industry and the specific processes involved. Industrial accidents can occur in various forms due to the complexity and diversity of industrial activities [2]. However, there are several types of industrial accidents that are relatively common across various sectors:

a) Falls
- Slips and Trips: Due to wet or uneven surfaces, leading to falls.
- Falls from Heights: Common in construction and manufacturing, often due to lack of proper fall protection.

b) Machinery Accidents
- Caught-in or Between Objects: Body parts or clothing getting caught in moving machinery or equipment.
- Struck by Objects: Being hit or struck by heavy objects or machinery parts.

c) Chemical Accidents
    - Spills and Leaks: Accidental release of hazardous chemicals, leading to exposure or environmental contamination.
    - Exposure: Contact with harmful chemicals, causing burns, respiratory issues, or other health problems.

d) Fire and Explosions
    - Industrial Fires: Due to faulty equipment, electrical issues, or chemical reactions.
    - Explosions: Resulting from combustible dust, gas leaks, or improper storage of explosive materials.

e) Electrical Accidents
    - Electric Shocks: Contact with live wires or faulty electrical equipment.
    - Electrocution: Severe electric shocks leading to fatal injuries.

f) Biological Hazards
    - Exposure to Pathogens: Common in healthcare and laboratory settings, leading to infections.
    - Animal Bites or Scratches: Relevant in agriculture or research involving animals.

g) Radiological Accidents
    - Exposure to Radiation: In nuclear facilities or healthcare, leading to radiation poisoning or long-term health issues.

h) Confined Space Accidents
    - Entrapment: Workers getting trapped in confined spaces with limited entry or exit points.
    - Hazardous Atmospheres: Exposure to toxic gases, lack of oxygen, or combustible environments in confined spaces.

i) Transportation Accidents
    - Vehicle Collisions: Involving trucks, forklifts, or other industrial vehicles.
    - Pedestrian Accidents: Workers being struck by moving vehicles within industrial premises.

j) Structural Failures
    - Building Collapses: Due to poor construction or maintenance.
    - Roof or Wall Failures: Resulting from heavy loads, weather, or aging structures.

k) Heat and Cold Stress
- Heat Exhaustion/Stroke: Caused by prolonged exposure to high temperatures.
- Hypothermia: Caused by exposure to extremely cold temperatures.

l) Noise-Related Issues
- Hearing Loss: Due to prolonged exposure to loud machinery or equipment.

m) Psychological Stress
- Work-Related Stress: Leading to mental health issues, burnout, or workplace accidents due to impaired concentration.

n) Manual Handling Injuries
- Strains and Sprains: Caused by lifting, pushing, or pulling heavy objects improperly.

o) Lack of Oxygen
- Asphyxiation: Occurs in confined spaces or areas with low oxygen levels.

13.6 Common Causes of Industrial Accidents

Preventing industrial accidents involves a combination of proper training, regular equipment maintenance, strict adherence to safety protocols, fostering a culture of safety within the organization, and continuous risk assessment and improvement of safety measures. Industrial accidents can occur due to various factors, often involving a combination of elements. Some common causes of industrial accidents are as follows [2, 3]:

a) Human Error
- Lack of Training: Inadequate training or lack of knowledge about operating machinery or handling hazardous materials.
- Fatigue: Tired workers are more prone to mistakes and accidents.
- Distractions: Inattentiveness due to distractions, such as personal issues or electronic devices, can lead to accidents.

b) Equipment Malfunction
   - Mechanical Failure: Malfunctions in machinery or equipment due to wear and tear, lack of maintenance, or manufacturing defects.
   - Incorrect Use: Misuse or improper operation of equipment can lead to accidents.
   - Lack of Safety Devices: Absence or malfunctioning of safety features like emergency stop buttons or guards.

c) Unsafe Working Conditions
   - Poor Housekeeping: Cluttered work areas, slippery floors, or inadequate lighting can contribute to accidents.
   - Inadequate Ventilation: Poor air quality due to lack of ventilation in areas with hazardous chemicals.
   - Lack of Personal Protective Equipment (PPE): Failure to provide or use appropriate safety gear like helmets, gloves, or goggles.

d) Hazardous Materials
   - Improper Storage: Incorrect storage of hazardous substances leading to leaks, spills, or reactions.
   - Lack of Training: Inadequate knowledge about handling and disposing of hazardous materials safely.
   - Chemical Reactions: Accidental mixing of chemicals that produce dangerous reactions.

e) Communication Issues
   - Lack of Communication: Insufficient communication between workers, especially during shift changes, can lead to misunderstandings and accidents.
   - Inadequate Signage: Absence of clear signs indicating hazards or emergency exits.

f) External Factors
   - Natural Disasters: Events like earthquakes, floods, or storms can damage industrial facilities.
   - Supplier Issues: Defective materials or parts supplied by external vendors.
   - Sabotage or Terrorism: Deliberate acts aimed at causing harm to the industrial facility.

g) Lack of Safety Protocols
   - Inadequate Safety Policies: Absence of comprehensive safety guidelines and protocols within the organization.
   - Failure to Follow Procedures: Disregarding established safety procedures and protocols.

h) Pressure to Meet Targets
   - Production Pressure: Workers may cut corners on safety protocols to meet production quotas or deadlines.

i) Inadequate Risk Assessment
   - Failure to Identify Risks: Incomplete or inadequate risk assessments that do not foresee potential hazards.

j) Inadequate Maintenance
   - Lack of Regular Maintenance: Failure to perform routine checks and maintenance on machinery and equipment.

13.7 Accident Investigation

Industrial accident investigation refers to the systematic process of examining the causes, events, and contributing factors of an industrial accident to determine what happened, why it happened, and how similar accidents can be prevented in the future. The primary goal of industrial accident investigation is to identify the root causes and underlying factors that led to the accident, rather than assigning blame to individuals. Each type of industrial accident requires specific safety measures and protocols to prevent them, emphasizing the importance of comprehensive training, proper equipment maintenance, and strict adherence to safety guidelines within industrial workplaces. The key components of an industrial accident investigation [2, 3]:

a) Immediate Response
   - Ensure Safety: The immediate priority is to ensure the safety of all personnel and secure the accident site to prevent further injuries or damage.
   - Preserve Evidence: Preserve the accident scene and collect relevant physical evidence before it is disturbed or tampered with.

b) Gather Information
   - Witness Interviews: Interview witnesses, including workers, supervisors, and anyone else who might have relevant information about the accident.
   - Documentation: Gather relevant documents, such as safety procedures, equipment manuals, and training records.

- Photographs and Videos: Capture images and videos of the accident site, equipment, and any relevant conditions.

c) Root Cause Analysis
- Identify Causes: Determine the immediate, underlying, and root causes of the accident. Immediate causes are the direct events leading to the accident, whereas underlying and root causes are the systemic issues that allowed the immediate causes to occur.
- Use Tools: Employ methods like the 5 Whys, Fishbone Diagrams (Ishikawa), or Fault Tree Analysis to analyze causative factors.

d) Report and Documentation
- Formal Report: Prepare a detailed report outlining the findings of the investigation, including the sequence of events, causes, and recommendations for prevention.
- Recommendations: Provide specific recommendations and corrective actions to prevent similar accidents in the future.
- Documentation: Maintain thorough records of the investigation process, findings, and actions taken.

e) Implement Corrective Actions
- Immediate Actions: Implement immediate corrective measures to address urgent safety concerns.
- Long-Term Solutions: Develop and implement long-term solutions to address systemic issues identified during the investigation.

f) Follow-Up and Monitoring
- Monitor Progress: Regularly follow up on the implementation of corrective actions to ensure they are effective and sustainable.
- Continuous Improvement: Use lessons learned from the investigation to improve safety protocols, training, and procedures within the organization.

g) Communication
- Internal Communication: Communicate the investigation findings and corrective actions to all relevant employees within the organization.
- External Communication: If necessary, communicate relevant information to regulatory authorities, insurance companies, or other external stakeholders.

Industrial accident investigations are critical for preventing future accidents, improving safety protocols, and fostering a culture of safety within industrial workplaces. They also contribute to legal and insurance processes, ensuring accountability and promoting overall workplace safety.

13.8 Importance of Conducting Accident Investigation

Conducting industrial accident investigations is essential for preventing future incidents, ensuring worker safety, complying with regulations, maintaining financial stability, fostering a positive workplace culture, promoting organizational learning, and enhancing community and public relations. Effective investigations are a cornerstone of a proactive approach to workplace safety. Conducting industrial accident investigations is crucial for several reasons, all of which contribute to enhancing workplace safety and preventing future accidents. Key reasons why investigating industrial accidents is important [3, 4]:

a) Preventing Future Incidents
   - Identify Root Causes: Investigations help determine the underlying causes of accidents, enabling organizations to address the root issues that led to the incident.
   - Implement Corrective Actions: Findings from investigations allow for the implementation of corrective actions and preventive measures to avoid similar accidents in the future.
   - Continuous Improvement: By learning from past mistakes, organizations can continuously improve their safety protocols and procedures, creating a safer work environment.

b) Ensuring Worker Safety
   - Protect Workers: Investigations help identify hazards and unsafe practices, allowing for immediate corrective actions to protect workers from potential harm.
   - Enhance Safety Training: Insights from investigations can be used to improve safety training programs, ensuring that workers are well-informed about potential risks and how to mitigate them.

c) Legal and Regulatory Compliance
    - Compliance: Proper investigations demonstrate an organization's commitment to complying with legal and regulatory requirements related to workplace safety and accident reporting.
    - Liability Management: Investigations provide accurate information that can be crucial in legal proceedings, insurance claims, and regulatory inspections. Proper documentation can mitigate legal liabilities.

d) Employee Morale and Trust
    - Employee Confidence: Thorough investigations show employees that their safety and well-being are taken seriously, boosting confidence in the organization's commitment to their welfare.
    - Trust Building: Transparent investigations build trust between management and employees, fostering a positive workplace culture.

e) Financial Impact
    - Reduced Costs: Preventing accidents through effective investigations reduces costs associated with medical expenses, compensation claims, legal fees, and potential fines or penalties.
    - Preservation of Assets: Proper investigations can prevent damage to equipment and facilities, preserving valuable assets and avoiding costly repairs or replacements.

f) Organizational Learning
    - Knowledge Sharing: Insights gained from accident investigations can be shared within the organization and even with industry peers, contributing to collective knowledge about safety best practices.
    - Preventing Recurrence: Learning from past accidents helps organizations avoid making the same mistakes, creating a safer workplace environment.

g) Community and Public Image
    - Community Relations: Demonstrating a commitment to safety through thorough investigations positively influences the organization's relationship with the community.
    - Public Image: A strong safety record, achieved in part through effective investigations, enhances the organization's public image and brand reputation.

13.9 Objectives of Accident Investigation

Industrial accident investigations play a crucial role in creating safer work environments, preventing injuries and fatalities, and fostering a culture of safety within industrial workplaces. The primary objectives of industrial accident investigation are to identify the root causes of the incident and to prevent similar accidents from occurring in the future. A thorough investigation is essential for several reasons:

a) Determine Causes
   - Immediate Causes: Identify the direct factors that led to the accident, such as equipment failure or human error.
   - Underlying Causes: Uncover systemic issues, organizational failures, or latent conditions contributing to the accident.

b) Prevention of Recurrence
   - Implement Corrective Actions: Develop and implement corrective measures addressing identified deficiencies to prevent similar incidents.
   - Proactive Measures: Introduce preventive actions to mitigate potential risks and improve overall safety protocols.

c) Legal and Regulatory Compliance
   - Liability Assessment: Determine liability and responsibility, crucial for legal and insurance purposes.
   - Compliance: Ensure compliance with local, national, and international regulations, standards, and codes related to workplace safety.

d) Improvement of Safety Protocols
   - Process Enhancement: Review existing safety protocols and procedures, and improve them based on the lessons learned from the accident.
   - Training and Awareness: Enhance training programs to educate employees about the incident, its causes, and preventive measures.

e) Accountability
   - Identify Responsible Parties: Determine if any individuals, departments, or contractors were at fault and hold them accountable.

- Cultural Change: Foster a culture of accountability and responsibility within the organization, encouraging safe practices at all levels.

f) Learning and Knowledge Sharing
- Lessons Learned: Extract valuable lessons from the incident and share them within the organization and, if applicable, with the industry to prevent similar accidents elsewhere.
- Best Practices: Promote the adoption of best practices and innovations in safety management based on investigation findings.

g) Employee and Public Safety
- Employee Welfare: Demonstrate the organization's commitment to employee safety, fostering trust and confidence among the workforce.
- Public Assurance: Ensure the surrounding community and the public that appropriate measures have been taken to prevent similar incidents, enhancing public safety.

h) Continuous Improvement
- Feedback Loop: Establish a feedback mechanism to continually assess the effectiveness of implemented safety measures and adjust policies as needed.
- Organizational Learning: Use investigation outcomes to improve risk management strategies and decision-making processes within the organization.

13.10 Structure of an Accident Report

An industrial accident report is a crucial document used to record detailed information about an accident that occurs in an industrial setting. It is important for the report to be accurate, objective, and thorough. The report should be reviewed by relevant parties, such as safety officers, managers, and legal advisors, to ensure completeness and accuracy before being filed and used for preventive actions and decision-making processes [4]. While the specific components of the report may vary based on the organization's policies and the nature of the incident, a comprehensive industrial accident report typically includes the following components:

1. Basic Information
    - Date and Time: When the accident occurred.
    - Location: Where the accident took place, including specific details about the department or area.
    - Weather Conditions: Note relevant weather conditions if they contributed to the incident.
2. Identification and Contact Information
    - Names and Roles: Names, job titles, and roles of the individuals involved, including witnesses.
    - Employee IDs: Unique identification numbers of the employees involved.
    - Contact Information: Phone numbers and addresses of the individuals involved.
3. Description of the Incident
    - Narrative: A detailed account of how the accident happened, including the sequence of events leading up to the incident.
    - Witness Statements: Accounts from witnesses, if available, regarding what they saw or heard.
    - Photographs: Pictures of the accident scene, equipment involved, or any relevant factors contributing to the incident.
4. Injury and Damage Details
    - Injuries: Description of injuries sustained by individuals involved, including medical diagnosis if available.
    - Property Damage: Details about damage to equipment, machinery, or other property.
    - Medical Treatment: Information about immediate medical care provided and the medical facility where the injured were taken.
5. Root Cause Analysis
    - Immediate Causes: Identify the direct factors that caused the accident.
    - Underlying Causes: Identify systemic or organizational factors contributing to the incident, such as lack of training, equipment malfunction, or inadequate safety protocols.
    - Contributing Factors: Identify any external or environmental factors that played a role in the accident.

6. Corrective Actions
    - Short-Term Actions: Immediate measures taken to address the situation, such as first aid, evacuations, or containment of hazards.
    - Long-Term Actions: Comprehensive measures and changes in procedures, equipment, or training to prevent similar incidents in the future.
7. Recommendations
    - Training Needs: Recommendations for additional training or retraining for employees involved or the organization as a whole.
    - Policy Changes: Suggestions for changes in company policies or safety procedures to prevent similar accidents.
    - Equipment Upgrades: Recommendations for upgrading or replacing faulty equipment to enhance safety.
8. Conclusion and Signatures
    - Summary: A concise summary of the incident, its causes, and the actions taken to prevent recurrence.
    - Signatures: Signatures of the individuals involved, including employees, witnesses, and supervisors, to confirm the accuracy of the report.
9. Appendix (Optional)
    - Supporting Documents: Include any additional documents, diagrams, or records relevant to the incident.

13.11 Steps for Conducting Accident Investigation

Conducting a thorough industrial accident investigation is crucial for understanding the causes of the incident and implementing measures to prevent future accidents. Organizations can conduct a thorough and effective industrial accident investigation, leading to a safer work environment and preventing similar incidents in the future [4]. The steps typically involved in an industrial accident investigation are the following:

i. Secure the Scene: Ensure the accident scene is safe for investigators to enter. Preserve the scene to prevent tampering or alteration of evidence.
ii. Provide Medical Care: Attend to injured individuals promptly, ensuring that they receive appropriate medical

attention. Document injuries and treatments for the investigation report.

iii. Notify Authorities: Comply with legal requirements by reporting the accident to relevant regulatory authorities.
iv. Form an Investigation Team: Assemble a multidisciplinary team including safety experts, engineers, and relevant personnel. Designate a team leader responsible for coordinating the investigation.
v. Gather Information: Collect witness statements, including those of employees, supervisors, and any other relevant parties. Document physical evidence, such as damaged equipment, tools, or materials. Review relevant documents, including safety procedures, training records, and equipment manuals.
vi. Conduct Interviews: Interview individuals involved in the accident as well as witnesses. Ask open-ended questions to gather detailed and accurate information.
vii. Examine Equipment and Systems: Inspect the machinery, equipment, and systems involved in the accident. Identify any defects, malfunctions, or maintenance issues.
viii. Review Procedures: Evaluate whether established safety procedures were followed. Identify any deviations from standard operating procedures (SOPs).
ix. Identify Root Causes: Analyze the information gathered to determine the immediate and underlying causes of the accident.
x. Develop Recommendations: Propose corrective actions to address identified deficiencies and prevent similar incidents. Prioritize recommendations based on their potential impact on safety.
xi. Create an Investigation Report: Document the investigation process, findings, and recommendations in a comprehensive report. Include witness statements, evidence, root cause analysis, and proposed corrective actions.
xii. Implement Corrective Actions: Ensure that the recommended corrective actions are implemented promptly. Monitor the progress of action items and verify their effectiveness.
xiii. Follow-Up and Review: Conduct follow-up inspections and interviews to ensure that implemented measures are

effective. Review the effectiveness of the corrective actions over time and make adjustments as necessary.

xiv. Communication: Communicate investigation findings and lessons learned to all relevant stakeholders. Use the incident as a learning opportunity to enhance safety awareness and training programs.

xv. Documentation and Record Keeping: Maintain detailed records of the investigation process, findings, and corrective actions taken. Use this information for future reference, training, and continuous improvement.

13.12 Different Methods of Accident Investigations

Accident investigation involves a systematic approach to understanding the causes and contributing factors of an incident. Various methods and techniques are used to conduct thorough and effective accident investigations. The choice of method or combination of methods depends on the nature of the incident, available resources, and the specific goals of the investigation. Employing a multidisciplinary approach and using appropriate methods can help investigators uncover the root causes and develop effective preventive measures. Some common methods employed in accident investigations are as follows:

1. Root Cause Analysis (RCA)
    - 5 Whys: Asking "why" repeatedly (typically five times) to drill down to the fundamental cause of the problem.
    - Fishbone Diagram (Ishikawa): Identifying potential causes by categorizing them into branches on a diagram, such as people, processes, equipment, materials, environment, and management.
2. Failure Mode and Effects Analysis (FMEA)
    - Systematic Assessment: Identifying potential failure modes of a system and evaluating their effects, causes, and detection mechanisms to prevent failures.
3. Event and Causal Factor Charting (ECFC)
    - Timeline Analysis: Creating a chronological chart of events leading up to the incident, including actions, decisions, and conditions, to identify causal factors.

4. Barrier Analysis
    - Barrier Identification: Identifying barriers (safety measures, procedures, and training) that should have prevented the accident and analyzing why they failed or were ineffective.
5. Change Analysis
    - Change Impact Assessment: Evaluating recent changes in procedures, personnel, equipment, or policies to assess their influence on the incident.
6. Witness Interviews
    - Structured Interviews: Conducting interviews with witnesses, victims, and involved personnel using predefined questions to gather firsthand accounts of the incident.
7. Data Analysis
    - Statistical Analysis: Analyzing incident data to identify patterns, trends, or correlations that can provide insights into the root causes.
8. Human Factors Analysis
    - Human Error Evaluation: Applying human factors principles to understand how human actions, cognition, and behavior contributed to the incident.
    - Cognitive Task Analysis: Investigating thought processes and decision-making strategies of individuals involved in the incident.
9. Accident Reconstruction
    - Engineering Analysis: Using engineering principles and simulations to reconstruct the sequence of events and understand the physical aspects of the incident.
10. Documentation and Records Review
    - Document Analysis: Reviewing relevant documents, procedures, training records, maintenance logs, and other documentation related to the incident.
11. Expert Consultation
    - Specialized Experts: Consulting experts in specific fields, such as engineering, safety, or human factors, to gain specialized knowledge relevant to the incident.
12. Legal and Regulatory Compliance Review
    - Compliance Assessment: Ensuring that the organization's practices comply with applicable laws, regulations, and industry standards.

13. Safety Culture Assessment
    - Organizational Culture Analysis: Evaluating the organization's safety culture, including leadership, communication, and employee attitudes toward safety.
14. Site Visits and Inspections
    - On-Site Assessment: Conducting physical inspections of the accident site, equipment, and relevant facilities to gather firsthand information.
15. Comparative Analysis
    - Benchmarking: Comparing the organization's safety practices and incident rates with industry benchmarks to identify areas for improvement.

13.13 Structure of an Accident Investigation Report

An accident investigation report is a crucial document that outlines the details of an accident, its causes, and the measures taken to prevent future incidents. It should be clear, concise, and comprehensive. Creating a well-structured and detailed accident investigation report is crucial for organizational learning and continuous improvement in safety management practices. A typical structure for an accident investigation report is as follows:

1. Title
    - "Accident Investigation Report" or a specific title related to the incident.
2. Introduction
    - Date, Time, and Location: When and where the accident occurred.
    - Description of Incident: A brief overview of what happened.
    - Purpose of the Report: Explain why the investigation was conducted and what the report aims to achieve.
3. Investigation Team
    - Names and Titles: List the names and job titles of individuals involved in the investigation.
4. Background Information
    - Context: Describe the environment, equipment, and conditions leading up to the accident.

- Involved Parties: Identify the people, equipment, and materials involved in the incident.

5. Detailed Incident Description
   - Sequence of Events: Provide a chronological account of what happened before, during, and after the accident.
   - Eyewitness Testimonies: Include statements from witnesses, victims, and anyone directly involved.
   - Photographs or Diagrams: Visual aids depicting the accident scene, equipment, and relevant details.
6. Immediate and Underlying Causes
   - Immediate Causes: Describe the direct factors that led to the accident.
   - Underlying Causes: Identify systemic or organizational issues contributing to the incident.
7. Root Cause Analysis
   - Method Used: Explain the technique (e.g., 5 Whys and Fishbone Diagram) used to identify the root causes.
   - Root Causes: Clearly state the fundamental reasons for the accident.
8. Corrective Actions
   - Specific Measures: Detail the actions taken to address both immediate and underlying causes.
   - Responsibilities: Assign responsibilities for implementing the corrective actions.
   - Timeline: Specify deadlines for the completion of each corrective action.
9. Preventive Measures
   - Proactive Steps: Describe measures to prevent similar incidents in the future.
   - Training and Education: Outline plans for educating employees about the incident and preventive measures.
10. Conclusion
    - Summary: Recap the key points of the investigation, including causes and actions taken.
    - Lessons Learned: Discuss lessons learned and how the organization plans to apply them.
11. Recommendations
    - Further Actions: Suggest additional measures or areas of improvement to enhance safety protocols.
    - Policy Changes: Propose changes to existing policies or procedures if necessary.

12. Appendices
    - Supporting Documents: Attach relevant documents, photographs, witness statements, or investigation forms.
    - References: Include references to standards, regulations, or internal policies used during the investigation.
13. Signatures and Approval
    - Investigators: Signatures of team members indicating their agreement with the report's findings.
    - Approvals: Signature of the responsible authority or management representative approving the report and its recommendations.
14. Distribution
    - List the individuals or departments receiving copies of the report for review and action.

13.14 Who Should Conduct the Accident Investigation?

The responsibility for conducting an accident investigation typically falls on trained individuals or a team within an organization. The exact composition of this team may vary based on the size of the organization, the nature of the accident, and industry regulations. The goal is to conduct a thorough, unbiased, and comprehensive investigation to identify root causes and prevent future incidents. Some guidelines on who should conduct the accident investigation are as follows:

1. Trained Safety Personnel
    - Safety Officers: Organizations often have dedicated safety officers or safety managers with specialized training in accident investigation.
    - Health and Safety Committees: In larger organizations, committees consisting of employees and safety experts might handle investigations.
2. Experienced Supervisors or Managers
    - Supervisors: Frontline supervisors who are familiar with the work processes and the individuals involved can be valuable investigators.

- Managers: Middle or senior managers might lead investigations, especially for significant incidents with widespread implications.

3. External Experts
   - Consultants: In complex situations, organizations might hire external safety consultants or experts to conduct impartial investigations.
   - Government Inspectors: Regulatory agencies may conduct their own investigations, especially for major accidents or in highly regulated industries.
4. Multidisciplinary Teams
   - Cross-Functional Teams: Including members from different departments (engineering, operations, safety, and maintenance) ensures a diverse perspective on the incident.
   - Human Resources: HR representatives can be involved, especially if the accident involves employee behavior or interactions.
5. Involvement of Workers
   - Employee Representatives: Inclusion of worker representatives or union members can enhance the investigation's credibility and fairness.
   - Witnesses: Employees who witnessed the incident are crucial sources of information and should be actively involved.
6. Consideration of Expertise
   - Technical Experts: For accidents involving specialized equipment or processes, individuals with technical expertise should be included in the team.
   - Legal Advisors: In situations where legal implications are significant, legal counsel might provide guidance.
7. Key Qualities
   - Objectivity: Investigators must remain impartial, focusing on facts rather than personal biases.
   - Analytical Skills: The ability to analyze data and draw logical conclusions is crucial.
   - Communication Skills: Investigators must effectively communicate with witnesses, employees, and management.

8. Involvement of Management
   - Management Oversight: While not directly involved in the investigation, higher management should oversee the process, ensuring impartiality and appropriate follow-up on recommendations.
   - Management Review: Investigative findings should be reviewed and approved by management before implementation of corrective actions.
9. Regulatory Compliance
   - Compliance Officers: If specific regulations or laws govern accident investigations in an industry, then compliance officers ensure that the investigation aligns with legal requirements.

13.15 What Should be Looked at as the Cause of an Accident?

When investigating the cause of an accident, it is important to adopt a systematic approach and consider a wide range of factors. It is possible to identify the multiple causes and contributory elements leading to an accident, enabling the implementation of effective preventive measures [5]. Key areas to examine when determining the cause of an accident:

1. Immediate Causes
   - Direct Actions: Identify the specific actions or events that directly led to the accident, such as a worker slipping or machinery malfunctioning.
   - Unsafe Acts: Determine if any individual involved committed an unsafe act, like not following procedures or operating equipment incorrectly.
   - Unsafe Conditions: Identify hazardous conditions like faulty equipment, inadequate lighting, or slippery floors that contributed to the accident.
2. Underlying Causes
   - Systems and Procedures: Analyze if existing systems, policies, or procedures were inadequate or not followed properly.

- Training and Competency: Assess the training provided to employees and whether they were adequately skilled and knowledgeable.
- Communication: Investigate if there was a breakdown in communication, leading to misunderstandings or lack of awareness.
- Supervision: Evaluate the supervision and management practices to ensure adequate oversight and guidance.
- Workplace Culture: Examine the organization's safety culture, including attitudes, values, and beliefs regarding safety within the workplace.

3. Root Causes
   - Management Systems: Investigate organizational processes and management systems, such as risk assessment, safety protocols, and monitoring mechanisms.
   - Organizational Culture: Assess the overall culture of the organization, including leadership commitment to safety and the willingness of employees to report concerns.
   - Resource Allocation: Examine if there were insufficient resources allocated to safety measures, including staffing, equipment, or training budgets.
4. Human Factors
   - Physical Factors: Consider human physical limitations, such as strength, vision, and hearing, which might have played a role in the accident.
   - Mental Factors: Evaluate mental factors like fatigue, stress, distraction, or lack of concentration that could have affected decision-making and reaction times.
5. External Factors
   - Environmental Conditions: Consider external factors like weather, lighting, or natural disasters that might have influenced the accident.
   - Supplier or Contractor Issues: Investigate if external parties, such as suppliers or contractors, were involved and if their actions or products contributed to the accident.
6. Equipment and Technology
   - Equipment Failure: Examine whether equipment malfunctioned due to wear, lack of maintenance, or manufacturing defects.

- Technological Failures: Investigate if technological systems or automation processes failed, leading to the accident.

7. Regulatory Compliance
   - Compliance with Regulations: Determine if the organization was compliant with relevant laws, regulations, and industry standards.
   - Permitting and Licensing: Check if necessary permits and licenses were in place and up-to-date.
8. Documentation and Records
   - Review Records: Analyze relevant records, such as maintenance logs, training records, and incident reports, for any patterns or discrepancies.
9. Witness Testimonies
   - Interview Witnesses: Gather statements from witnesses, victims, and others involved to understand their perspectives and insights.
10. Previous Incidents
    - Historical Data: Consider any past incidents or near-misses that might reveal recurring issues or patterns within the organization.

13.16 Fact Collection for Industrial Accident Investigations

Collecting accurate and comprehensive facts is crucial in any industrial accident investigation. Proper data collection helps in understanding the incident, identifying causes, and formulating effective corrective and preventive measures. Investigators can gather accurate and reliable facts, enabling a thorough analysis of the industrial accident and the formulation of appropriate preventive measures [4]. Facts are typically collected for industrial accident investigations in the following ways:

- Initial Incident Report: Begin with the initial incident report, which often provides basic details like time, date, location, and a brief description of the accident. This report might be filed by the employees, supervisors, or witnesses who first respond to the incident.

- Witness Interviews: Interview witnesses, including employees, supervisors, and any other individuals present during the incident. Their firsthand accounts can provide critical information about the sequence of events and conditions leading up to the accident.
- Victim Statements: Interview the individuals directly involved in the accident. Their statements can provide insights into their actions, the equipment or processes involved, and any factors that might have contributed to the incident.
- Scene Examination: Visit the accident site as soon as possible to examine the physical surroundings. Document the layout, equipment, machinery, tools, materials, and any relevant environmental factors. Take photographs and videos to capture the scene from various angles.
- Documentation Review: Review relevant documents, such as work orders, SOPs, safety protocols, training records, equipment manuals, and maintenance logs. These documents can provide information about standard procedures and whether they were followed.
- Equipment Inspection: Inspect the equipment involved in the accident to identify any malfunctions, defects, or signs of wear and tear. Document serial numbers, maintenance records, and any visible damage.
- Data Collection Tools: Use measurement tools such as tape measures, thermometers, or gauges to collect specific data related to the accident, such as distances, temperatures, pressures, or other relevant metrics.
- Expert Consultation: Consult experts or specialists if necessary, especially in cases involving complex machinery, hazardous materials, or specific technical knowledge.
- Records and CCTV Footage: Access records, including surveillance camera footage, to reconstruct the events leading to the accident. CCTV footage can provide valuable visual information.
- Employee Statements and Surveys: Conduct surveys or anonymous interviews with employees to gather information about workplace conditions, safety culture, and any concerns they might have.
- Regulatory Compliance: Ensure compliance with regulatory requirements. Investigate whether the organization was

following applicable laws, standards, and regulations at the time of the accident.

- Recreation of Events: Where applicable, recreate the sequence of events using simulations, computer models, or reenactments to understand the dynamics of the accident.
- Timeline Construction: Create a detailed timeline of events leading up to, during, and after the accident. A chronological order of actions can help in identifying causative factors.
- Documentation Preservation: Safeguard all collected evidence, statements, photographs, and records for future reference, legal purposes, or external investigations.

13.17 What Should be Done if the Accidental Investigation Reveals Human Error?

If an industrial accident investigation reveals human error as one of the contributing factors, then it is essential to approach the situation with a focus on prevention and learning rather than blame. Addressing human errors in an industrial setting requires a multifaceted approach that involves improving processes, enhancing training, supporting employees, and fostering a culture of safety. By focusing on prevention and continuous improvement, organizations can minimize the likelihood of similar errors in the future. The steps to be considered for such problem are the following [6]:

1. Avoid Blame
    - Non-Punitive Approach: Emphasize a non-punitive approach to encourage reporting of errors without fear of reprisal.
    - Focus on Systems: Recognize that human errors often occur due to systemic issues rather than individual shortcomings.
2. Root Cause Analysis
    - Identify Underlying Causes: Determine why the human error occurred. Was it due to lack of training, unclear procedures, fatigue, or other factors?
    - Use Human Factors Analysis: Apply human factors principles to understand the interaction between humans and the systems they work with.

3. Training and Education
    - Provide Training: Offer additional training or retraining to employees involved in the error, emphasizing the correct procedures and best practices.
    - Continuous Learning: Implement ongoing training programs to enhance skills and knowledge, addressing potential areas of weakness.
4. Improve Procedures and Processes
    - Clear Procedures: Ensure that procedures are well-documented, easy to understand, and followed consistently.
    - Simplify Processes: Streamline complex processes to minimize the likelihood of errors.
    - Standardize Practices: Implement standardized procedures to reduce variability and improve predictability.
5. Implement Redundancies and Safety Checks
    - Redundant Systems: Introduce redundant systems or safety measures to catch errors before they escalate.
    - Checklists and Cross-Verification: Implement checklists and cross-verification processes to catch potential mistakes.
6. Address Fatigue and Stress
    - Work Hours: Evaluate working hours, shift patterns, and workload to prevent fatigue-related errors.
    - Mental Health Support: Provide resources and support for mental health to help employees cope with stress.
7. Promote a Safety Culture
    - Open Communication: Encourage open communication about errors and near-misses to learn from them.
    - Leadership Role: Leadership should set an example by admitting mistakes, emphasizing safety, and encouraging reporting.
8. Monitor and Evaluate
    - Continuous Monitoring: Continuously monitor processes and employee performance to identify patterns and address potential issues.
    - Data Analysis: Use data analytics to identify trends and areas for improvement related to human errors.

9. Feedback and Recognition
   - Feedback Mechanism: Establish a feedback mechanism where employees can provide input on processes, identifying potential areas of improvement.
   - Recognition: Acknowledge and recognize employees for identifying errors and suggesting improvements, fostering a proactive safety culture.
10. Document Lessons Learned
    - Share Findings: Share the findings of the investigation, focusing on lessons learned and preventive measures, promoting organizational learning.
11. Legal and Regulatory Compliance
    - Compliance Measures: Ensure that corrective actions align with legal and regulatory requirements to avoid future violations.

References

1. Ferry, T.S., *Modern accident investigation and analysis*, John Wiley & Sons, United States, 1988.
2. Katsakiori, P., Sakellaropoulos, G., Manatakis, E., Towards an evaluation of accident investigation methods in terms of their alignment with accident causation models. *Saf. Sci.*, 47, 7, 1007–1015, 2009.
3. Sklet, S., Comparison of some selected methods for accident investigation. *J. Hazard. Mater.*, 111, 1-3, 29–37, 2004.
4. Lundberg, J., Rollenhagen, C., Hollnagel, E., What-You-Look-For-Is-What-You-Find–The consequences of underlying accident models in eight accident investigation manuals. *Saf. Sci.*, 47, 10, 1297–1311, 2009.
5. Kletz, T.A., Accident investigation: Keep asking "why?". *J. Hazard. Mater.*, 130, 1-2, 69–75, 2006.
6. Shappell, S.A. and Wiegmann, D.A., A human error approach to accident investigation: The taxonomy of unsafe operations. *Int. J. Aviat. Psychol.*, 7, 4, 269–291, 1997.

14
Risk-Based Decision-Making

Abstract
This chapter explains the fundamental ideas required for effective risk-based decision-making. It also highlights the relevant data qualities required to allow data utilization to make risk-based choices and discusses the handling of information issues. It also discusses the importance of risk-based decision-making, different methods, and classifications. It also discusses the advantages and disadvantages of risk-based decision-making, their application, and effective ways of using risk-based decision-making.

Keywords: Risk identification, risk assessment, risk management

14.1 Background

Decisions are established based on the necessity of making any decision, whether to do something or not, or for choosing one alternative among a set of options. Options are often limited by societal, technological, safety, commercial, and environmental restrictions. Effective decision-making requires a grasp of these parameters, their relative importance, evaluation possibilities, and choosing the best option. Risk management's major purpose is to use safety-related data to identify and control the possible effects of risks in the aviation system before an accident or significant event happens. The classification of safety data using standard taxonomies allows the data to be seen in more dimensions to spot dangers more effectively, making risk management much more effective.

14.2 Introduction

Before making assessments, risk-based decision-making is a strategic technique that entails examining the risks and uncertainties associated with various possibilities. This approach is prevalent in business, project

Abhishek Barua, Kanchan Kumari, Sumit Saha, Siddharth Jeet and Swastik Pradhan. Institutional and Industrial Safety Engineering Practices, (335–352) © 2025 Scrivener Publishing LLC

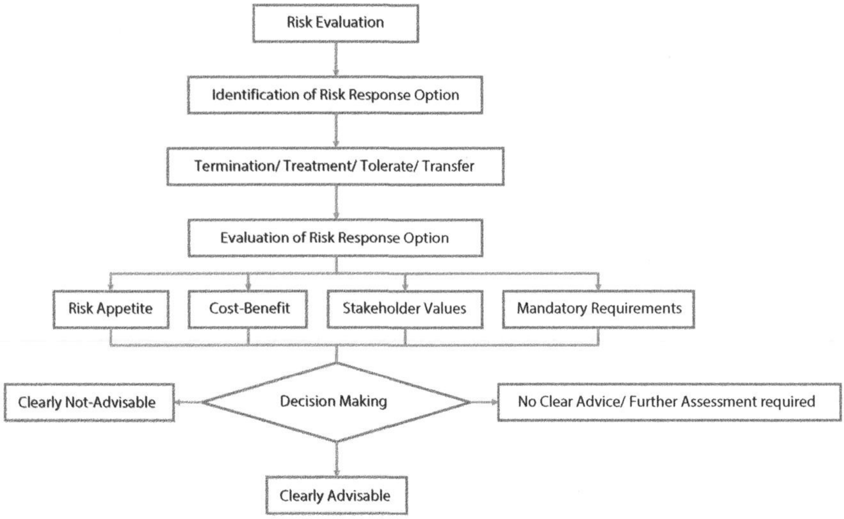

Figure 14.1 Structure of risk-based decision-making.

management, and other industries where decision-makers must negotiate unpredictable settings. The general decision-making steps remain unchanged in risk-based decision-making: identify the concerns, analyze the options, and implement the decision. Risk-based decision-making is distinguished by the fact that it is founded on a systematic understanding of the risk-reward ratio and uncertainty [1]. Figure 14.1 depicts the overall framework of risk-based decision-making. It has been divided into five sections for ease of comprehension.

14.2.1 Setting the Target

The first phase in this decision-making process is to identify and define a set of group target. This, like every other phase, should be a collaborative endeavor confirmed by group consensus. The engagement of stakeholders in identifying and addressing issue areas is not illustrated here, although it is an important phase in the process. Many advantages arise with increasing group engagement, such as higher acceptance of the objectives (resulting in improved motivation) and better comprehension of the target (resulting in enhanced capacity to effectively support goal-related actions). While target selection is critical, it is covered in other guidelines and regulations and will not be covered here. What remains is the construction of the remainder of the process, in which evaluations and strategies for achieving those objectives are developed, as well as the establishment of a feedback loop that allows for constant enhancement of the targets.

14.2.2 Risk Assessment

Risk assessment detects the possible hazards present in the system and ranks them (and/or their components) in terms of the previously stated risk criteria. As such, it aims to address issues such as:

- What is possible that could go wrong?
- How likely is it that anything will go wrong?
- What are the impacts?

14.2.3 Risk Management

A risk management action strategy may be created once a screened and prioritized list of hazards has been created. Because risk mitigation strategies may fluctuate greatly depending on the scenario, no exhaustive list of alternative management activities is conceivable. In general, risk management seeks to address issues such as the following:

- What exactly can be done?
- What solutions are accessible, and what are the related choices?
- How will current decisions affect future options?

14.2.4 Assessment of Impact

An evaluation of the effectiveness of the counter-measures deployed is required to offer input for any risk assessments, objective setting, and selection. For identifying and rating the variations in risk arising from risk management events, both objective and subjective techniques should be employed, as with other data collection and investigation in this practice. For objective data, look into governmental and industrial databases for relevant and reliable information.

14.2.5 Risk Communication

An evaluation of the effectiveness of the countermeasures deployed is required for offering input for future risk assessments, objective setting, and selection. For identifying and rating the variations in risk arising from risk management events, both objective and subjective techniques should be employed, as with other data collection and investigation in this practice. For objective data, look into governmental and industrial databases for relevant and reliable information. Finally, outcomes will be reported to stakeholders and up through the chain of command. Senior commanders would go over these to get information for their individual risk assessment and control efforts.

14.3 Steps Involved in Risk-Based Decision-Making

Risk-based decision-making (RBDM) is a structured approach to decision-making that takes into account the potential risks associated with various options. By following these steps, organizations can make informed decisions that take into account the potential risks and increase the likelihood of achieving their objectives while managing uncertainties effectively [2]. The steps involved in RBDM typically include the following:

a) Define the Decision Context:
   - Clearly articulate the decision that needs to be made.
   - Identify the goals and objectives associated with the decision.

b) Identify and Define Risks:
   - Systematically identify and list potential risks that may impact the decision.
   - Categorize risks based on their nature (e.g., financial, operational, and reputational).

c) Assess Risks:
   - Evaluation of the chance and severity of each risk that is identified.
   - Prioritization of risks centered on their potential impact and probability of occurrence.

d) Risk Tolerance and Acceptance:
   - Define the organization's or project's risk tolerance level.
   - Determine which risks are acceptable and which ones are not, considering the defined risk tolerance.

e) Generate Decision Alternatives:
   - Identify and generate potential alternatives or options for the decision.

- Consider the potential impact of each alternative on the identified risks.

f) Evaluate Alternatives:
- Assess each alternative in terms of its potential benefits and drawbacks.
- Consider how each alternative addresses or mitigates the identified risks.

g) Decision Analysis:
- Use decision analysis techniques to evaluate and compare the alternatives.
- Consider the trade-offs between risks and benefits.

h) Select Preferred Alternative:
- Choose the alternative based on the evaluation that best fits the organization's objectives.
- Ensure that the selected alternative aligns with the established risk tolerance.

i) Implement Decision:
- Develop an implementation plan for the chosen alternative.
- Consider risk mitigation strategies during the implementation phase.

j) Monitor and Adapt:
- Establish a monitoring and review process to track the outcomes of the decision.
- Be prepared to adapt the decision or its implementation based on new information or changes in the risk landscape.

k) Document and Communicate:
- Document the decision-making process, including the identified risks, assessments, and the rationale for the chosen alternative.
- Communicate the decision and its associated risks to relevant stakeholders.

l) Review and Learn:
- After the decision has been implemented and the outcomes are observed, conduct a post-implementation review.
- Identify lessons learned and use them to improve future decision-making processes.

14.4 Importance of Risk-Based Decision-Making

Risk-based decision-making is crucial in various fields and industries because it allows individuals and organizations to make informed choices that take into account potential uncertainties and adverse outcomes. Risk-based decision-making is not just a risk mitigation strategy; it is a fundamental approach that can lead to more resilient, adaptive, and successful organizations across various sectors [3]. Some key reasons why RBDM is important are as follows:

- Optimized Resource Allocation: By assessing and prioritizing risks, resources can be allocated more efficiently. This ensures that efforts are focused on mitigating or managing the most critical risks, leading to better overall outcomes.
- Improved Decision Quality: Understanding and considering risks in decision-making processes enhances the quality of decisions. It helps in anticipating potential challenges and developing strategies to address them, leading to more successful outcomes.
- Enhanced Problem Identification: A risk-based approach encourages a thorough analysis of potential issues and uncertainties. This facilitates the identification of problems early in the decision-making process, allowing for timely and effective mitigation strategies.
- Better Strategic Planning: Organizations can develop more robust strategic plans by incorporating risk assessments. This helps in aligning goals with potential challenges and opportunities, making the strategy more adaptable and resilient.
- Regulatory Compliance: In many industries, compliance with regulations and standards is essential. Risk-based decision-making helps in identifying and addressing potential compliance risks, reducing the likelihood of legal and regulatory issues.
- Innovation and Growth: Taking calculated risks is often necessary for innovation and growth. A risk-based decision-making process allows organizations to identify and assess risks associated with new ventures or opportunities, enabling them to make informed decisions that balance potential rewards with potential drawbacks.

- Crisis Preparedness: Understanding and addressing risks in advance helps organizations to be better prepared for potential crises. This preparedness can minimize the impact of unexpected events and improve the organization's ability to respond effectively.
- Continuous Improvement: Regularly reviewing and updating risk assessments fosters a culture of continuous improvement that refines risk management strategies and adapts to changing circumstances.
- Cost Savings: Proactively managing risks can result in cost savings by avoiding or minimizing the impact of adverse events. This includes potential financial losses, legal liabilities, and damage to reputation.

14.5 Classification of Risk in Perspective of Risk-Based Decision-Making

Risk-based decision-making involves assessing and managing uncertainties and potential consequences to make informed choices. The classification of risk in the context of decision-making typically involves categorizing risks based on various criteria. The importance of each category may vary depending on the nature of the organization, industry, and external environment [4]. Effective RBDM involves identifying, assessing, prioritizing, and mitigating risks to achieve organizational objectives. General classification of risks in the context of RBDM is as follows:

a) Financial Risk:
    - Market Risk: Arises from changes in market conditions such as interest rates, exchange rates, and commodity prices.
    - Credit Risk: The risk of financial loss due to the failure of a borrower or counterparty to meet their financial obligations.

b) Operational Risk:
    - Internal Processes: Risks associated with the efficiency and reliability of internal processes, systems, and people.
    - External Events: Risks arising from external events such as natural disasters, political instability, or technological failures.

- Human Factor: Risks related to human actions, including errors, fraud, or inadequate training.

c) Compliance Risk:
- Legal and Regulatory Compliance: Risks associated with failure to comply with laws and regulations governing the business.
- Ethical Compliance: Risks related to ethical standards and corporate governance.

d) Strategic Risk:
- Reputation Risk: The risk of damage to an organization's reputation, often resulting from negative public perception.
- Competitive Risk: Risks associated with changes in the competitive landscape or failure to adapt to market trends.

e) Information Technology (IT) Risk:
- Cybersecurity Risk: The risk of unauthorized access, data breaches, or disruptions to IT systems.
- Data Privacy Risk: Risks associated with the collection, storage, and use of sensitive information.

f) Supply Chain Risk:
- Supplier Risks: Risks related to the reliability, financial stability, and performance of suppliers.
- Logistics Risks: Risks associated with transportation, distribution, and delivery of goods and services.

g) Environmental and Sustainability Risk:
- Environmental Impact: Risks related to environmental factors, including climate change, resource scarcity, and pollution.
- Social and Governance (ESG) Risks: Risks associated with social responsibility, diversity and inclusion, and ethical business practices.

h) Project Risk:
- Scope and Schedule: Risks related to project planning and execution.
- Budget and Cost: Risks associated with financial aspects of project management.

i) Reputational Risk:
- Brand Image: Risks related to how the organization is perceived by customers, investors, and the public.

- Crisis Management: Risks associated with responding to and managing crises effectively.

j) Country or Geopolitical Risk:
- Political Instability: Risks associated with changes in political leadership, policies, and geopolitical events.
- Economic Conditions: Risks related to economic instability, inflation, and currency exchange rates.

14.6 Different Types of Risk-Based Decision-Making

Risk-based decision-making involves making decisions while taking into account the potential risks and uncertainties associated with different options. There are various types of risk-based decision-making approaches used in different fields. The application of risk-based decision-making can vary across industries and contexts. The common theme is the consideration of risks and uncertainties in the decision-making process to achieve more informed and resilient outcomes. Common types of RBDM are the following:

a) Financial Risk-Based Decision-Making
- Credit Risk: Assessing the risk of default on loans or financial obligations.
- Market Risk: Evaluating the potential losses due to fluctuations in market conditions.
- Operational Risk: Considering risks related to internal processes, systems, people, and external events.

b) Project Risk-Based Decision-Making
- Risk Assessment: Identifying and analyzing potential risks that may impact project objectives.
- Risk Mitigation: Developing strategies to reduce or eliminate the impact of identified risks.
- Contingency Planning: Planning for unforeseen events and having backup strategies in place.

c) Healthcare Risk-Based Decision-Making
- Clinical Decision-Making: Balancing the benefits and risks of different treatment options for patients.
- Risk Management in Healthcare: Identifying and managing risks related to patient safety, regulatory compliance, and other factors.

d) Environmental Risk-Based Decision-Making
   - Environmental Impact Assessment: Assessing the potential environmental risks and impacts of projects or activities.
   - Sustainable Decision-Making: Considering the long-term environmental consequences of decisions.

e) Information Technology (IT) Risk-Based Decision-Making
   - Cybersecurity Risk Management: Assessing and managing risks related to information security and data breaches.
   - IT Governance: Making decisions to align IT strategies with business objectives while managing associated risks.

f) Supply Chain Risk-Based Decision-Making
   - Supply Chain Risk Assessment: Identifying and assessing risks associated with the supply chain, such as disruptions, delays, and quality issues.
   - Risk Mitigation in Supply Chain: Developing strategies to minimize the impact of potential risks on the supply chain.

g) Regulatory Compliance Risk-Based Decision-Making
   - Compliance Risk Management: Making decisions to ensure compliance with laws, regulations, and industry standards.
   - Ethical Decision-Making: Considering ethical considerations and potential reputational risks associated with business decisions.

h) Strategic Risk-Based Decision-Making
   - Strategic Planning: Integrating risk considerations into the overall strategic planning process.
   - Scenario Planning: Considering different future scenarios and making decisions that are robust across a range of possibilities.

i) Operational Risk-Based Decision-Making
   - Business Continuity Planning: Planning for and making decisions to ensure the continuity of business operations in the face of unexpected events.
   - Quality Management: Making decisions to ensure the quality and reliability of products or services.

14.7 Advantages of Risk-Based Decision-Making

Risk-based decision-making (RBDM) is an approach that involves assessing and prioritizing decisions based on the potential risks involved. Adopting a risk-based decision-making approach can lead to more effective, efficient, and resilient decision-making processes, contributing to the overall success and sustainability of an organization. There are several advantages to adopting a risk-based decision-making framework:

a) Efficient Resource Allocation: RBDM allows organizations to allocate resources more efficiently by focusing on high-impact decisions that carry significant risks. Resources can be directed toward areas where the potential consequences of a decision, whether positive or negative, are most substantial.
b) Improved Decision Quality: By systematically considering and evaluating risks, decision-makers can make more informed and better-quality decisions. Anticipating potential risks allows for the development of mitigation strategies, leading to more robust and resilient decisions.
c) Enhanced Risk Awareness: RBDM encourages a proactive approach to risk management by promoting awareness of potential risks associated with various decisions. Decision-makers become more attuned to the uncertainties and vulnerabilities associated with their choices.
d) Adaptability to Changing Environments: RBDM equips organizations with the ability to adapt to changing circumstances. By continually assessing risks, decision-makers can adjust their strategies and actions in response to new information or evolving conditions.
e) Regulatory Compliance: In certain industries, regulatory requirements mandate the consideration of risks in decision-making processes. Adopting RBDM ensures compliance with such regulations and helps organizations avoid legal issues.
f) Early Identification of Opportunities: RBDM not only focuses on risks but also on potential opportunities. By considering both sides, organizations can identify opportunities for innovation, growth, and competitive advantage.
g) Cost Reduction: Proactively managing risks through RBDM can help prevent costly mistakes and unforeseen events. This

can lead to cost savings by avoiding unnecessary expenditures related to poor decisions.

h) Continuous Improvement: RBDM is an iterative process that allows for continuous improvement. Organizations can learn from the outcomes of decisions, refine their risk assessments, and adjust their decision-making processes accordingly.

i) Strategic Alignment: RBDM helps align decision-making with organizational objectives and strategies. By prioritizing decisions based on their impact on strategic goals, organizations can ensure that resources are directed toward achieving overarching objectives.

14.8 Disadvantages of Risk-Based Decision-Making

Risk-based decision-making, like any approach, has its own set of disadvantages and challenges. It is important to note that, while RBDM has its challenges, it can be a valuable approach when implemented thoughtfully and in conjunction with a robust risk management framework. Regular review and adjustment of the risk management process can help mitigate some of these disadvantages. Some potential drawbacks are as follows:

- Subjectivity and Bias: Risk assessments can be subjective, relying on the judgment of individuals or groups. This can introduce biases based on personal experiences, perspectives, or organizational culture.
- Incomplete Information: In some cases, it might be challenging to obtain all the necessary information to make an accurate risk assessment. Incomplete data can lead to inaccurate risk evaluations and decisions.
- Overlooking Rare Events: RBDM tends to focus on more common and foreseeable risks. Rare or unpredictable events, often referred to as "black swan" events, may be overlooked, leading to inadequate preparation for such occurrences.
- Complexity and Time-Consuming: Performing a thorough risk assessment can be time-consuming and complex. This might be a disadvantage in fast-paced industries or situations where quick decisions are required.

- Difficulty in Quantifying Risks: Assigning precise quantitative values to risks can be challenging, as it often involves making assumptions. This can lead to disagreements among stakeholders and difficulties in comparing and prioritizing risks.
- Dynamic Nature of Risks: Risks are dynamic and can change over time. A risk identified today may evolve or become obsolete tomorrow. Continuous monitoring and reassessment are necessary, which can be resource-intensive.
- Resistance to Change: Implementing a risk-based approach may face resistance within organizations, especially if it requires a significant shift in the existing decision-making processes. Employees and stakeholders may resist changes they perceive as time-consuming or disruptive.
- Overemphasis on Compliance: In some cases, organizations may focus excessively on complying with regulations rather than addressing the actual risks. This can result in a false sense of security and neglect of critical areas.
- False Sense of Security: Relying solely on risk assessments may create a false sense of security, especially if the assessments are not updated regularly. Organizations may become complacent and fail to adapt to new emerging risks.
- Inadequate Consideration of Interconnected Risks: Risks are often interconnected, and changes in one area can impact others. Failing to consider these interdependencies can lead to unexpected consequences and ineffective risk management.

14.9 Applications of Risk-Based Decision-Making

Risk-Based Decision-Making (RBDM) is a systematic approach to making decisions in the presence of uncertainty. It involves identifying, assessing, and prioritizing risks to inform decision-making processes. RBDM is applied across various industries and fields to enhance decision-making and improve outcomes. RBDM provides a structured framework for decision-makers to systematically evaluate and address uncertainties, resulting in more informed and effective decision-making. Here are some key applications of RBDM:

1. Project Management
    - Risk Assessment: Identifying potential risks and uncertainties associated with project objectives, timelines, and budget.
    - Risk Prioritization: Prioritizing risks based on their potential impact and likelihood, enabling the allocation of resources to manage high-priority risks.
2. Financial Decision-Making
    - Investment Decisions: Evaluating the risks associated with various investment options to make informed financial decisions.
    - Portfolio Management: Managing investment portfolios by considering the risk-return trade-off for different assets.
3. Healthcare
    - Patient Safety: Assessing and managing risks in healthcare processes to enhance patient safety and reduce adverse events.
    - Clinical Trials: Identifying and managing risks associated with the development and testing of new medical treatments.
4. Environmental Management
    - Environmental Impact Assessment: Evaluating and managing the environmental risks associated with projects or activities.
    - Sustainable Development: Making decisions that balance economic, social, and environmental considerations by considering associated risks.
5. Information Technology (IT) and Cybersecurity
    - Data Security: Assessing and mitigating risks related to the security and privacy of sensitive data.
    - Cybersecurity Risk Management: Identifying and managing risks associated with cyber threats and vulnerabilities.
6. Supply Chain Management
    - Supply Chain Resilience: Evaluating and managing risks in the supply chain to ensure continuity and responsiveness.
    - Vendor Risk Management: Assessing and mitigating risks associated with third-party suppliers and partners.

7. Regulatory Compliance
    - Compliance Management: Ensuring compliance with regulations by identifying and managing risks associated with non-compliance.
    - Legal Decision-Making: Evaluating legal risks and uncertainties to inform legal strategies and decisions.
8. Energy Sector
    - Project Risk Analysis: Assessing risks in energy projects, such as oil and gas exploration or renewable energy development.
    - Safety Management: Identifying and mitigating risks associated with the production and distribution of energy resources.
9. Insurance and Risk Management
    - Underwriting Decisions: Assessing risks associated with insuring individuals, businesses, or assets.
    - Claims Management: Evaluating and managing risks when processing insurance claims.
10. Emergency Response Planning
    - Disaster Preparedness: Identifying and planning for potential risks and disasters to enhance emergency response capabilities.
    - Crisis Management: Implementing strategies to manage risks during crises and emergencies.

14.10 How to Make Risk-Based Decision-Making More Effective?

Risk-based decision-making is a systematic approach that involves assessing potential risks and making decisions based on the likelihood and impact of those risks. By following few steps and integrating risk-based decision-making principles into an organizational culture, one can enhance the effectiveness of decision-making processes and increase the likelihood of achieving successful outcomes.

Some key steps and strategies to make RBDM effective are the following:

a) Identify and Assess Risks
    - Clearly identify and define potential risks associated with the decision at hand.

- Assess the likelihood and potential impact of each identified risk.
- Consider both internal and external factors that could contribute to the risks.

b) Prioritize Risks
- Prioritize risks based on their potential impact and likelihood of occurrence.
- Focus on addressing high-priority risks that could have a significant impact on the decision or project.

c) Gather Information
- Collect relevant data and information to support the risk assessment.
- Consult with subject matter experts and stakeholders to gain diverse perspectives.

d) Establish Risk Tolerance
- Determine the organization's or project's risk tolerance level.
- Define acceptable levels of risk for different aspects of the decision.

e) Develop Mitigation Strategies
- Identify and implement strategies to mitigate or reduce high-priority risks.
- Consider preventive measures to avoid risks and corrective measures to address risks that have already occurred.

f) Cost-Benefit Analysis
- Evaluate the cost and benefits of different decision options, considering the potential risks.
- Assess the trade-offs between risk mitigation measures and the overall benefits of the decision.

g) Communication and Transparency
- Communicate the identified risks and proposed mitigation strategies to relevant stakeholders.
- Foster an open and transparent communication culture to ensure that everyone is informed about the risks involved.

h) Monitor and Review
- Continuously monitor the risk landscape throughout the decision-making process.
- Regularly review and update risk assessments based on new information or changes in the environment.

i) Learn from Experience
   - Document and analyze the outcomes of previous decisions to learn from both successes and failures.
   - Use lessons learned to improve future RBDM.
j) Continuous Improvement
   - Implement a continuous improvement process for risk management.
   - Regularly review and update risk management processes and tools based on feedback and experience.
k) Training and Skill Development
   - Provide training for decision-makers and team members on risk management principles and techniques.
   - Develop the skills needed to effectively assess, communicate, and manage risks.

References

1. Peterman, R.M. and Anderson, J.L., Decision analysis: a method for taking uncertainties into account in risk-based decision making. *Hum. Ecol. Risk Assess.: Int. J.*, 5, 2, 231–244, 1999.
2. St. Onge, J.R. and Floresco, S.B., Prefrontal cortical contribution to risk-based decision making. *Cereb. Cortex*, 20, 8, 1816–1828, 2010.
3. Bohnenblust, H. and Slovic, P., Integrating technical analysis and public values in risk-based decision making. *Reliab. Eng. Syst. Saf.*, 59, 1, 151–159, 1998.
4. Maybee, B., Lowen, S., Dunn, P., Risk-based decision making within strategic mine planning. *Int. J. Min. Miner. Eng.*, 2, 1, 44–58, 2010.

15

Risk-Based Maintenance

Abstract
This chapter explains the fundamental ideas required for effective risk-based maintenance. It also highlights the relevant data qualities required to allow data utilization to make risk-based choices and discusses the handling of information issues. It also discusses the importance of risk-based maintenance, different methods, and classifications. It also discusses the advantages and disadvantages of risk-based maintenance, their application, and effective ways of using risk-based maintenance.

Keywords: Risk identification, risk assessment, maintenance

15.1 Background

Engineers realize that if they do not account for a little bit of interruption for maintenance and repairs, then they are likely to wind up losing a significant amount of time, revenue, and efforts when their technology, infrastructure, or machinery fails. As a result, understanding how risk-based maintenance methodologies and applications might improve the maintenance plan can be advantageous. Maintenance is essential in every organization, whether it is providing time and space for staff to relax or shutting down large factories for simple housekeeping. The ultimate purpose of maintenance management is to reduce unexpected equipment breakdowns while being cost-effective. There is no flawless maintenance plan that is applicable to everyone. This opens the door to a variety of techniques, such as maintenance based on conditions, reliability-centered maintenance, and overall productivity maintenance. Risk-based maintenance is a method of allocating assets for maintenance based on an asset's risk of failure. Risk-based thinking in facility management prioritizes the resources with the highest potential for failure. These assets are more often checked and maintained than others. The model reduces the risk of breakdown in the most economically feasible way by determining which assets to repair and when.

Abhishek Barua, Kanchan Kumari, Sumit Saha, Siddharth Jeet and Swastik Pradhan. Institutional and Industrial Safety Engineering Practices, (353–366) © 2025 Scrivener Publishing LLC

15.2 Introduction

A risk-based maintenance strategy prioritizes maintenance tasks depending on the amount of risk related to infrastructure or assets. Risk-based maintenance aims to maximize maintenance resources by concentrating attention on vulnerable elements, structures, or operations. It provides a technique to assist businesses in determining the risk worth of their properties so as to make socially prudent resource selections [1]. Adopting a risk-based maintenance approach assists facilities in reducing costly breakdowns. This approach helps organizations minimize the likelihood of equipment failure, reduce downtime, and enhance overall reliability [2]. Risk-based maintenance is a dynamic and proactive approach that aligns maintenance efforts with organizational goals, enhances equipment reliability, and contributes to overall operational efficiency. In addition, the results of risk-based maintenance protect employees, assets, and the company as a whole [3].

15.3 Importance of Risk-Based Maintenance

Risk-based maintenance is a proactive approach to maintenance that focuses on managing assets based on their risk profiles. It involves assessing the likelihood and consequences of equipment failures and prioritizing maintenance activities accordingly. The importance of risk-based maintenance lies in its ability to enhance cost-effectiveness, improve asset reliability, prioritize safety, and contribute to overall operational excellence. By aligning maintenance activities with the risks associated with assets, organizations can achieve a more sustainable and efficient approach to managing their physical infrastructure [4]. The importance of risk-based maintenance can be understood through several key aspects:

- Cost Optimization: Risk-based maintenance helps optimize maintenance costs by directing resources toward critical assets that are more prone to failure or have higher consequences of failure. This avoids unnecessary and costly maintenance on less critical equipment.
- Asset Reliability: By targeting maintenance efforts on high-risk assets, risk-based maintenance enhances overall asset reliability. This ensures that critical equipment is available and operational when needed, minimizing downtime and improving productivity.

- Safety Improvement: Prioritizing maintenance based on risk helps identify and address safety-critical issues promptly. This proactive approach contributes to a safer working environment by reducing the likelihood of equipment failures that could lead to accidents or injuries.
- Operational Efficiency: Risk-based maintenance enables organizations to focus on maintaining equipment that has the most significant impact on operations. This enhances efficiency by minimizing disruptions and optimizing the performance of critical assets.
- Strategic Planning: Risk-based maintenance facilitates strategic planning by providing insights into the health and performance of assets. This information is crucial for long-term planning, capital budgeting and decision-making related to asset management.
- Regulatory Compliance: Many industries are subject to regulations regarding equipment maintenance and safety. Risk-based maintenance helps organizations comply with these regulations by providing a structured and risk-informed approach to maintenance practices.
- Data-Driven Decision-Making: Risk-based maintenance relies on data and analytics to assess risks and prioritize maintenance activities. This data-driven approach enables more informed decision-making, allowing organizations to continuously improve their maintenance strategies based on performance feedback.
- Life Cycle Management: Risk-based maintenance supports effective life cycle management of assets by considering their condition, performance, and potential failure risks. This helps organizations make decisions about repairs, upgrades, or replacements at the right time.
- Resource Allocation: Risk-based maintenance allows organizations to allocate resources efficiently by focusing on the areas that need immediate attention. This prevents over-maintenance of low-risk assets and ensures that resources are directed where they are most needed.
- Predictive Maintenance: Risk-based maintenance often integrates with predictive maintenance technologies, such as condition monitoring and sensor data, to anticipate and address potential issues before they lead to failures. This predictive approach minimizes unplanned downtime.

15.4 How to Conduct Risk-Based Maintenance

Conducting risk-based maintenance involves integrating risk assessments into maintenance decision-making processes to prioritize and optimize maintenance activities [5]. Here is a systematic guide on how to conduct risk-based maintenance:

1. Outline the Objectives and Scope: Clearly state the objectives of the maintenance program and the scope of assets to be considered. Understanding the goals and boundaries is essential for a focused risk-based maintenance approach.
2. Identify Critical Assets: Identify and prioritize critical assets that have a significant impact on overall operations, safety, or financial performance. Consider factors such as criticality, reliability, and importance to the organization's mission.
3. Risk Identification: Systematically identify potential risks associated with the critical assets. This involves considering failure modes, historical performance data, industry best practices, and input from relevant stakeholders.
4. Risk Assessment: Evaluate the possibility and effects of the identified risks. To assign risk scores, use either qualitative or quantitative methodologies. This stage aids in risk prioritization based on criticality and probable impact on asset performance.
5. Prioritize Maintenance Activities: Based on the risk assessment, prioritize maintenance activities for critical assets. Focus on high-priority risks that require immediate attention. This step ensures that resources are allocated where they are most needed.
6. Select Maintenance Strategies: Choose appropriate maintenance strategies for each critical asset based on the risk assessment. Maintenance strategies may include preventive maintenance, predictive maintenance, condition-based maintenance, or a combination of these. Tailor the strategies to address the specific risks identified.
7. Implementation of Maintenance Plans: Implement the selected maintenance plans and strategies. This involves scheduling maintenance activities, allocating resources, and ensuring that the identified risks are effectively managed through proactive maintenance efforts.

8. Continuous Monitoring: Establish a monitoring system to track the performance of critical assets and the effectiveness of the chosen maintenance strategies. Regularly assess whether the implemented strategies are reducing the identified risks.
9. Feedback and Improvement: Collect feedback from maintenance activities and use it to continuously improve the risk-based maintenance process. This may involve adjusting maintenance strategies, revisiting risk assessments, or updating priorities based on evolving conditions.
10. Documentation and Reporting: Document the entire risk-based maintenance process, including risk assessments, maintenance plans, and performance monitoring results. Generate reports to communicate the status of critical assets and the effectiveness of maintenance strategies to relevant stakeholders.
11. Training and Communication: Ensure that relevant personnel are trained on the risk-based maintenance process and understand their roles in its implementation. Effective communication is crucial for the success of risk-based maintenance, so keep stakeholders informed about the progress and outcomes.
12. Adaptation to Changes: Be adaptable to changes in the operating environment, technology, or organizational goals. Regularly review and update the risk-based maintenance process to ensure its alignment with the evolving needs of the organization.

The risk-based maintenance approach is implemented for every system in the structure. A high-pressure vessel, for example, is an example of a system. This system will be surrounded by systems that transport fluid between and within the vessel. The system's potential failure mechanisms are first identified. Then, for each risk, a conventional risk-based maintenance strategy is used. Figure 15.1 depicts the risk-based maintenance framework. Important steps are as follows:

a) Collection of Data: Data must be gathered for each identified risk. This comprises information regarding the risk, its broad implications, and general approaches for mitigating and forecasting the risk.

Figure 15.1 Framework of risk-based maintenance.

b) Evaluation of Risk: Both the likelihood of the risk and the impact of the risk are estimated in the surroundings of the facility under assessment during the risk appraisal stage.
c) Ranking of Risks: When the risk assessment is finished, the likelihood and consequence are added together to determine the overall risk. This overall risk is compared to pre-determined risk levels. As a consequence, the risk might be classified as either satisfactory or unfavorable.
d) Creation of Inspection Plan: Whenever the risk is deemed intolerable, a strategy is developed to check the structure using a particular monitoring technique. Alternatively, if it is more cost-effective and technically viable, then a preventable maintenance program may be used.
e) Proposal for Mitigating Measures: At this point, a concept for risk mitigation based on status surveillance and upkeep is being developed.
f) Reassessment: In the end, the idea is assessed in light of other variables, including legal and administrative needs. If the suggestion's requirements are not satisfied, then the procedure is restarted. If not, then the maintenance suggestion is implemented.

15.5 Advantages of Risk-Based Maintenance

Risk-based maintenance offers several advantages for organizations across various industries. It offers a holistic and proactive approach to asset management, aligning maintenance efforts with the actual risks associated with equipment [6]. This results in cost savings, improved reliability, and enhanced overall operational performance for organizations. Some key advantages of implementing a risk-based approach to maintenance are as follows:

- Cost Efficiency: Risk-based maintenance enables organizations to allocate resources more efficiently by focusing maintenance efforts on critical assets with higher risks of failure. This helps reduce unnecessary maintenance costs on low-risk equipment.
- Improved Asset Reliability: By prioritizing maintenance activities based on risk, organizations can enhance the reliability of critical assets. This leads to improved operational performance and reduces the likelihood of unexpected downtime.
- Enhanced Safety: Risk-based maintenance allows organizations to identify and address safety-critical issues proactively. By focusing on assets with higher safety risks, organizations can minimize the chances of accidents or failures that could compromise employee safety.
- Optimized Maintenance Schedules: Risk-based maintenance helps organizations develop more effective maintenance schedules. By considering the risk profiles of assets, maintenance activities can be scheduled at optimal times, reducing disruptions to operations.
- Better Decision-Making: The risk-based approach provides decision-makers with valuable insights into the criticality of assets. This information supports more informed decisions related to repairs, upgrades, replacements, and overall asset management strategies.
- Extended Equipment Life: Through targeted and proactive maintenance, risk-based maintenance can contribute to extending the life of equipment. This is achieved by addressing potential issues before they lead to major failures or degradation.

- Compliance with Regulations: Many industries have regulations and standards related to equipment maintenance. Risk-based maintenance helps organizations comply with these requirements by providing a structured and systematic approach to maintenance practices.
- Strategic Planning: Risk-based maintenance contributes to long-term strategic planning by offering insights into the health and performance of assets. This information is valuable for budgeting, forecasting, and making decisions about capital investments.
- Predictive Maintenance Integration: Risk-based maintenance often integrates with predictive maintenance technologies, allowing organizations to use real-time data and condition monitoring to predict and address potential issues before they escalate.
- Resource Optimization: Risk-based maintenance helps organizations optimize their human and financial resources by directing efforts toward areas that truly need attention. This prevents over-maintenance of low-risk assets and ensures that resources are allocated where they have the most impact.
- Adaptability to Changing Conditions: Risk-based maintenance is adaptable to changing conditions and evolving risks. As the business environment or technology changes, organizations can adjust their risk assessments and maintenance strategies accordingly.
- Improved Overall Equipment Effectiveness: By focusing on critical assets, risk-based maintenance contributes to improving overall equipment effectiveness, which is a key metric for measuring the efficiency and effectiveness of production processes.

15.6 Disadvantages of Risk-Based Maintenance

While risk-based maintenance offers numerous advantages, it is essential to acknowledge the potential disadvantages and challenges associated with its implementation. It is important to address them proactively to maximize the benefits of risk-based maintenance. Customizing the implementation to suit the specific needs and context of the organization can help mitigate some of these challenges [7]. Some of the disadvantages include the following:

- Complexity and Expertise: Implementing risk-based maintenance may require a sophisticated understanding of both the equipment being maintained and the risk assessment process. It may be challenging for organizations without the necessary expertise or resources to carry out comprehensive risk assessments.
- Initial Implementation Costs: The initial setup of risk-based maintenance systems, including the acquisition of data collection tools, sensors, and specialized software, can be costly. Small or resource-constrained organizations may find it challenging to make this upfront investment.
- Data Quality and Availability: Risk-based maintenance relies heavily on accurate and up-to-date data. In situations where data is lacking or of poor quality, the effectiveness of the risk assessment can be compromised. Ensuring data quality and availability can be resource-intensive.
- Resistance to Change: Implementing risk-based maintenance may require a cultural shift within an organization. Employees accustomed to traditional maintenance practices may resist changes in procedures or mindset, affecting the successful adoption of risk-based maintenance.
- Overemphasis on Technical Aspects: Risk-based maintenance systems often focus on technical aspects of risk, potentially overlooking other critical factors, such as organizational culture, human factors, or external influences, which can also affect asset reliability.
- False Sense of Security: Relying solely on a risk-based approach may lead to a false sense of security. While it prioritizes maintenance efforts based on risk, it does not eliminate the possibility of unexpected failures or events.
- Continuous Monitoring Challenges: Continuous monitoring of equipment conditions is crucial for effective risk-based maintenance. However, assets, implementing continuous monitoring systems can be logistically challenging and expensive.
- Modeling Uncertainties: The models used in risk assessment may have uncertainties. If these uncertainties are not

properly considered and managed, then they can introduce inaccuracies into the risk analysis, affecting decision-making.

- Resource Allocation Dilemmas: While risk-based maintenance aims to optimize resource allocation by focusing on high-risk assets, there is a risk of neglecting lower-priority assets. This may lead to deferred maintenance, resulting in unexpected failures.
- Incompatibility With Legacy Systems: Introducing risk-based maintenance may be difficult in organizations with existing legacy maintenance systems that are not compatible with a risk-based approach. Integration challenges can hinder the smooth adoption of risk-based maintenance.
- Regulatory Compliance Challenges: Depending on the industry, regulatory requirements may not align seamlessly with a risk-based approach. Adhering to specific regulations while implementing risk-based maintenance can be challenging and may require careful navigation.

15.7 Application of Risk-Based Maintenance in Different Areas

Risk-based maintenance is a valuable approach for a variety of industries and organizations where asset management is crucial. In essence, any industry or organization that relies on critical assets to achieve its objectives can benefit from risk-based maintenance [8]. Risk-based maintenance provides a structured and proactive approach to maintenance, helping organizations prioritize efforts, allocate resources efficiently, and reduce the overall risk of equipment failures. Some sectors and contexts where risk-based maintenance is particularly relevant are as follows:

- Manufacturing Industries: Risk-based maintenance is widely used in manufacturing industries where the failure of critical equipment can result in production downtime, quality issues, and significant financial losses.
- Oil and Gas Industry: In the oil and gas sector, where the failure of equipment can lead to environmental disasters, production delays, and safety hazards, risk-based maintenance is essential for managing and mitigating risks.

- Power Generation and Utilities: Power plants and utilities rely on critical infrastructure to generate and distribute electricity. Risk-based maintenance helps in optimizing maintenance efforts to ensure reliable and continuous power supply.
- Aerospace and Aviation: In the aerospace industry, where safety is paramount, risk-based maintenance is crucial for ensuring the reliability of aircraft and minimizing the risk of failures that could lead to accidents.
- Transportation and Logistics: In the transportation sector, risk-based maintenance is used to manage the maintenance of vehicles, railway systems, and other critical infrastructure, ensuring the safety and reliability of transportation networks.
- Healthcare: In healthcare, where equipment reliability is essential for patient safety and effective medical treatments, risk-based maintenance is applied to manage and prioritize maintenance activities.
- Chemical and Process Industries: Industries dealing with chemical processes often involve complex and critical equipment. Risk-based maintenance is employed to minimize the risks of equipment failure, which can have severe safety and environmental consequences.
- Mining and Extractive Industries: In mining and extractive operations, where equipment failure can result in production delays, safety hazards, and environmental issues, risk-based maintenance helps optimize maintenance practices.
- Infrastructure Management: Risk-based maintenance is relevant in managing the maintenance of critical infrastructure such as bridges, tunnels, and dams. It ensures the safety and reliability of these structures.
- Information Technology (IT) and Data Centers: In the IT sector, particularly in data centers, risk-based maintenance is applied to manage the maintenance of servers, cooling systems, and other infrastructure to prevent disruptions and downtime.
- Facility Management: Organizations responsible for managing large facilities, such as commercial buildings, stadiums, or shopping centers, use risk-based maintenance to optimize maintenance activities and ensure the reliability of building systems.

- Water and Wastewater Treatment: In the water and wastewater treatment industry, risk-based maintenance helps manage the maintenance of pumps, valves, and treatment processes to ensure the continuous and safe supply of water.

15.8 How to Make Risk-Based Maintenance More Efficient?

Making risk-based maintenance more efficient involves optimizing the processes involved in identifying, assessing, and managing risks to achieve better outcomes with fewer resources. Organizations can streamline their risk-based maintenance processes, enhance decision-making, and achieve a more efficient and effective approach to asset management. Some strategies to enhance the efficiency of risk-based maintenance are as follows:

- Data-Driven Decision Making: Utilize data and analytics to inform decision-making. Implement predictive maintenance technologies, condition monitoring, and reliability-centered maintenance strategies to collect and analyze data on asset performance. This enables more accurate risk assessments and targeted maintenance interventions.
- Prioritize Critical Assets: Focus on critical assets that have a high impact on operations or safety. By prioritizing maintenance efforts on the most crucial components, you can maximize the impact of your resources and reduce the risk of major disruptions.
- Continuous Risk Monitoring: Implement continuous monitoring systems to track the health and performance of assets in real-time. This allows for early detection of potential issues, enabling proactive maintenance before failures occur.
- Integration with Maintenance Systems: Integrate risk-based maintenance processes with existing computerized maintenance management systems or enterprise asset management systems. This integration streamlines data management, facilitates workflow automation, and enhances overall efficiency.
- Risk-Based Inspection: Implement risk-based inspection methodologies for equipment. This involves assessing the risk associated with equipment failures and tailoring

inspection plans accordingly. RBI optimizes inspection frequency and resources based on risk profiles.

- Optimized Maintenance Scheduling: Develop a dynamic maintenance schedule that takes into account the current risk levels of assets. Adjust maintenance activities based on real-time risk assessments and prioritize tasks according to the evolving risk landscape.
- Cross-Functional Collaboration: Foster collaboration between different departments, such as maintenance, operations, and reliability engineering. This ensures a holistic approach to risk management and facilitates the sharing of insights and expertise.
- Training and Skill Development: Invest in training and skill development for maintenance personnel. Equip them with the knowledge and tools needed to effectively implement risk-based maintenance strategies. Well-trained staff can make quicker and more accurate decisions, improving overall efficiency.
- Performance Metrics and Key Performance Indicators: Establish and monitor performance metrics and key performance indicators to evaluate the effectiveness of your risk-based maintenance program. This includes measuring the impact on equipment reliability, maintenance costs, and overall operational efficiency.
- Continuous Improvement: Regularly review and assess the effectiveness of your risk-based maintenance program. Identify areas for improvement and adjust strategies accordingly. This continuous improvement mindset ensures that your maintenance processes remain aligned with organizational goals.
- Risk Communication: Improve communication channels regarding risk information. Ensure that relevant stakeholders are aware of the identified risks, mitigation strategies, and the rationale behind maintenance decisions. Clear communication fosters a shared understanding and commitment to risk management goals.
- Technology Adoption: Explore and adopt emerging technologies such as the Internet of Things, artificial intelligence, and machine learning. These technologies can enhance the efficiency of data collection, analysis, and decision-making in the risk-based maintenance process.

References

1. Arunraj, N.S. and Maiti, J., Risk-based maintenance—Techniques and applications. *J. Hazard. Mater.*, 142, 3, 653–661, 2007.
2. Khan, F.I. and Haddara, M.M., Risk-based maintenance (RBM): a quantitative approach for maintenance/inspection scheduling and planning. *J. Loss Prev. Process Ind.*, 16, 6, 561–573, 2003.
3. Khan, F.I. and Haddara, M., Risk-based maintenance (RBM): A new approach for process plant inspection and maintenance. *Process Saf. Prog.*, 23, 4, 252–265, 2004.
4. Krishnasamy, L., Khan, F., Haddara, M., Development of a risk-based maintenance (RBM) strategy for a power-generating plant. *J. Loss Prev. Process Ind.*, 18, 2, 69–81, 2005.
5. Sakai, S., Risk-based maintenance, vol. 17, JR East Technical Review, 2010.
6. Özcan, E., Yumuşak, R., Eren, T., Risk based maintenance in the hydroelectric power plants. *Energies*, 12, 8, 1502, 2019.
7. Dey, P.K., Ogunlana, S.O., Naksuksakul, S., Risk-based maintenance model for offshore oil and gas pipelines: a case study. *J. Qual. Maint. Eng.*, 10, 3, 169–183, 2004.
8. Cullum, J., Binns, J., Lonsdale, M., Abbassi, R., Garaniya, V., Risk-Based Maintenance Scheduling with application to naval vessels and ships. *Ocean Eng.*, 148, 476–485, 2018.

16

Safety Key Performance Indicators

Abstract

This chapter explains the fundamental ideas required for effective use of safety key performance indicators. It also highlights the relevant steps to define and track the safety key performance indicators. It also discusses large varieties of safety key performance indicators along with their uses and consequences. It also discusses the advantages and disadvantages of safety key performance indicators, their application, and effective ways of using safety key performance indicators.

Keywords: Hazard identification, safety, safety key performance indicators

16.1 Background

Safety management is a paramount concern for organizations across various industries. It is critical to understand a few safety issues in order to make continual enhancements to an industry's health and safety. To effectively monitor and improve safety performance, safety key performance indicators (or safety KPIs) are vital in this process because they allow one to make evaluations and position them in relation to all of the measures one have set in place. This allows one to identify flaws in the organization's existing occupational safety systems as well as share their triumphs. Safety KPI in safety management offer insights into the effectiveness of safety programs, help in preventing accidents, and support continuous improvement [1].

16.2 Introduction

Safety key performance indicators (safety KPIs) are metrics used to measure and evaluate the success of safety programs, practices, and total safety performance within an organization. These indicators help in monitoring

Abhishek Barua, Kanchan Kumari, Sumit Saha, Siddharth Jeet and Swastik Pradhan. Institutional and Industrial Safety Engineering Practices, (367–400) © 2025 Scrivener Publishing LLC

progress, identifying areas for improvement, and ensuring a safe working environment. Monitoring health and safety KPIs enable a firm to establish how safe the workplace is for employees and if the organization is in compliance with regulations. Monitoring these key performance indicators assists businesses in determining which safety requirements they are already meeting and which ones they may need to devote additional resources to meeting [2].

In the sphere of safety and health, safety KPIs are classified as lagging (reactive, following) or leading (proactive, forerunning). When picking KPIs, it is crucial to strike a good balance between leading and lagging indicators, just as it is when selecting proper techniques and objectives for occupational safety tasks [3].

a) Leading Indicators: These are proactive metrics that help predict and prevent accidents or incidents. They are also known as "preventive indicators" because they are not focused on past events but rather on actions that can be taken or current circumstances that may be changed in order to avoid future safety issues. They are forward-looking and focus on activities and conditions that have the potential to impact safety. Examples of leading indicators in safety management may include the number of safety training hours completed, the frequency of safety inspections, or the percentage of employees reporting near-miss incidents.
b) Lagging Indicators: These are reactive metrics that provide information about past safety performance. They examine the efficiency of present levels of occupational safety objectively and in contrast to other firms by measuring the consequences of safety activities. These indicators depend on data collected on safety incidents and consequences—events that happened and cannot be easily undone. Examples of lagging indicators include the number of recordable injuries, lost workdays due to accidents, or the total cost of workers' compensation claims. They are labeled as "lagging" because they are preoccupied with the past. Still, lagging indicators lack any guidance on how to get better.

A balanced approach to safety management frequently involves the use of both indicators that are leading and lagging. Leading indicators assist organizations in identifying and addressing possible safety concerns before they cause accidents, whereas lagging indicators give historical safety performance and may be used for benchmarking and compliance reporting [4].

16.3 Defining and Tracking Safety Key Performance Indicators

Defining and tracking safety KPIs is crucial for assessing and improving the safety performance of an organization. Safety KPIs are measurable values that provide insights into various aspects of safety management, helping organizations monitor their progress, identify areas for improvement, and ensure a safe working environment. By defining, tracking, and continuously improving safety KPIs, organizations can foster a safety culture, reduce the frequency and severity of incidents, and create a workplace environment that prioritizes the well-being of employees. Regularly reassessing and refining safety KPIs ensures that they remain aligned with organizational goals and industry best practices [5].

16.3.1 Defining of Safety KPIs

- Identify Objectives: Clearly outline the safety objectives of the organization. These objectives should support with the overall mission and values, emphasizing the importance of employee well-being, accident prevention, and compliance with safety regulations.
- Engage Stakeholders: Involve relevant stakeholders, including safety professionals, management, and frontline workers, in the process of identifying and defining safety KPIs. Consider their perspectives to ensure a comprehensive and inclusive approach.
- Select Appropriate Safety KPI: Choose safety KPIs that are relevant to the industry, organization size, and specific safety goals.
- Set Targets: Establish realistic and achievable targets for each safety KPI. Targets should be measurable, time-bound, and aligned with the organization's overall safety objectives. These targets serve as benchmarks for performance.

16.3.2 Tracking of Safety KPIs

- Collect Data: Implement systems and processes to collect relevant data for each safety KPI. This may involve incident reporting systems, safety inspections, training records, and other sources of safety-related information.
- Regular Monitoring: Set up a regular monitoring schedule to track safety KPIs. Depending on the nature of the KPIs, monitoring may occur daily, weekly, monthly, or quarterly. Consistent monitoring helps detect trends and assess the effectiveness of safety initiatives.
- Data Analysis: Analyze the collected data to gain insights into the organization's safety performance. Look for patterns, identify root causes of incidents, and assess the impact of safety interventions. Data analysis is crucial for making informed decisions.
- Report and Communicate: Prepare regular reports summarizing safety KPI performance. Communicate these reports to relevant stakeholders, including employees, managers, and safety committees. Transparency in reporting fosters a culture of accountability and continuous improvement.
- Review and Evaluate: Regularly review safety KPIs and evaluate whether the organization is meeting its targets. If targets are not being met, then conduct a thorough analysis to understand the reasons behind the performance gaps and implement corrective actions.
- Adjust Targets as Needed: Based on ongoing performance reviews and evaluations, adjust targets as needed. If the organization consistently meets or exceeds targets, then consider raising the bar to drive continuous improvement.
- Continuous Improvement: Use the perceptions gained from tracking safety KPIs to drive constant improvement in safety management systems. Implement lessons learned from incidents, identify opportunities for training and awareness, and refine safety procedures accordingly.
- Benchmarking: Compare the safety KPIs with industry benchmarks or standards. Benchmarking provides context for understanding performance relative to peers and can help identify areas where additional improvements are needed.

16.3.3 Some Common Safety KPIs

Safety key performance indicators (KPIs) are metrics used to measure and evaluate the effectiveness of safety programs and initiatives in an industrial setting. These indicators help organizations monitor their safety performance, identify areas for improvement, and track progress over time [6]. These safety KPIs provide a comprehensive overview of an organization's safety performance, covering aspects of injury frequency, severity, compliance, and proactive safety measures. Regularly tracking and analyzing these indicators can guide continuous improvement efforts in industrial safety. Some common safety KPIs in industrial safety are listed below.

16.3.3.1 Lost Time Injury Frequency Rate (LTIFR)

Lost time injury frequency rate (LTIFR) is a key safety performance indicator that is widely used in various industries to assess the frequency of workplace injuries that result in lost workdays. It is a crucial metric for monitoring and evaluating the safety performance of an organization over a specific period. LTIFR is calculated as the number of lost time injuries per one million hours worked. The formula for LTIFR is as follows:

$$LTIFR = \left(\frac{Number\ of\ Lost\ Time\ Injuries}{Total\ Hours\ Worked} \right) \times 1{,}000{,}000 \quad (16.1)$$

where:

- Number of Lost Time Injuries: This includes work-related injuries or illnesses that result in an employee being unable to perform their regular job duties for a specific period. Lost time injuries often involve medical treatment, restricted work activity, or days away from work.
- Total Hours Worked: This refers to the total number of hours worked by all employees during the specified time frame, typically expressed in millions of hours to align with the standard LTIFR unit.
- 1,000,000: This factor is included to standardize the rate per one million hours worked, making it easier to compare safety performance across different organizations and time periods.

Interpreting the LTIFR: A lower LTIFR indicates a safer work environment, as it implies fewer lost time injuries per million hours worked. A higher LTIFR suggests a higher frequency of lost time injuries relative to the total hours worked, indicating a potential need for improved safety measures.

Organizations often track LTIFR over specific time intervals (e.g., monthly, quarterly, or annually) to assess trends and identify areas for improvement in their safety programs. It is essential to complement LTIFR with other safety KPIs and contextual information to gain a comprehensive understanding of an organization's safety performance.

16.3.3.2 Total Recordable Injury Rate (TRIR)

The total recordable injury rate (TRIR) is another critical safety performance indicator used in various industries to assess the overall safety performance of an organization. TRIR measures the total number of recordable injuries per one million hours worked. It provides a broader perspective on workplace safety by considering injuries that may not result in lost workdays but still require medical treatment or other forms of restricted work activity. The formula for TRIR is as follows:

$$TRIR = \left(\frac{Number\ of\ Recordable\ Injuries}{Total\ Hours\ Worked} \right) \times 1{,}000{,}000 \quad (16.2)$$

where:

- Number of Recordable Injuries: This includes all work-related injuries or illnesses that result in medical treatment beyond first aid, restricted work activity, days away from work, or other recordable cases as defined by the relevant safety regulations.
- Total Hours Worked: Similar to LTIFR, this refers to the total number of hours worked by all employees during the specified time frame, typically expressed in millions of hours.
- 1,000,000: This factor is included to standardize the rate per one million hours worked, making it easier to compare safety performance across different organizations and time periods.

Interpreting the TRIR: Like LTIFR, a lower TRIR indicates a safer work environment, suggesting fewer recordable injuries per million hours worked. A higher TRIR implies a higher frequency of recordable injuries relative to the total hours worked, highlighting potential areas for improvement in safety programs.

It is important to note that TRIR provides a more inclusive view of workplace injuries compared to LTIFR, as it considers a broader range of incidents. Regular monitoring of TRIR over time helps organizations track trends, assess the effectiveness of safety programs, and implement targeted interventions to enhance workplace safety.

16.3.3.3 Days Away, Restricted, or Transferred (DART) Rate

The DART rate is another important safety performance indicator used to assess workplace safety. It specifically focuses on cases where employees are unable to perform their regular job duties due to work-related injuries or illnesses. The DART rate is calculated as the number of cases involving days away from work, restricted work activity, or job transfer per 100 full-time employees. The formula for DART Rate is as follows:

$$DART = \left(\frac{Number\,of\,DART\,Cases}{Total\,Hours\,Worked\,by\,All\,Employees} \right) \times 200{,}000 \quad (16.3)$$

where:

- Number of DART Cases: This includes cases where employees are away from work, have restricted work activity, or are transferred to another job due to work-related injuries or illnesses.
- Total Hours Worked by All Employees: This is the total number of hours worked by all employees during the specified time frame.
- 200,000: This factor is included to standardize the rate per 100 full-time employees, making it easier to compare safety performance across different organizations and time periods.

Interpreting the DART Rate: A lower DART rate indicates a safer work environment, suggesting fewer cases of employees being away from work, having restricted work activity, or being transferred due to work-related injuries or illnesses. A higher DART rate implies a higher frequency of cases where employees are away from work or have restricted work activity relative to the total hours worked, signaling potential areas for improvement in safety programs.

Similar to other safety performance indicators, organizations should use the DART rate in conjunction with other relevant metrics and contextual information to gain a comprehensive understanding of their safety performance. Regular monitoring of the DART rate helps organizations track trends, identify areas for improvement, and implement targeted safety interventions to enhance workplace safety.

16.3.3.4 Near-Miss Reporting Rate

The near-miss reporting rate is a safety performance indicator that measures the frequency of reported near misses in the workplace. A near miss is an unanticipated incident that may have caused harm, disease, or damage but did not. Tracking near misses is crucial for proactive safety management, as it provides insights into potential hazards and allows organizations to take preventive measures before incidents escalate. The near-miss reporting rate is calculated as the number of reported near misses per unit of measurement (e.g., per 1,000 hours worked or per month). The formula is as follows:

$$\textit{Near Miss Reporting Rate} = \left(\frac{\textit{Number of Reported Near Misses}}{\textit{Total Hours Worked or Other Relevant Unit}}\right) \times \textit{Multiplier} \quad (16.4)$$

where:

- Number of Reported Near Misses: This includes all near misses that have been reported by employees. Encouraging a culture of reporting near misses is essential for capturing valuable information about potential hazards.
- Total Hours Worked or Other Relevant Unit: This represents the total hours worked by all employees during the specified time frame or another relevant unit of measurement.
- Multiplier: This factor is included to standardize the rate based on the chosen unit of measurement (e.g., per 1,000 hours worked). The multiplier ensures consistency when comparing rates across different organizations or time periods.

Interpreting the Near-Miss Reporting Rate: A higher near-miss reporting rate suggests a proactive safety culture where employees are actively identifying and reporting potential hazards. It indicates that the organization encourages open communication about safety concerns. A lower near-miss reporting rate may suggest a need for improvement in the reporting culture. Organizations should encourage and educate employees on the importance of reporting near misses to prevent future incidents.

Monitoring and analyzing the near-miss reporting rate can help organizations identify trends, assess the efficiency of their safety programs, and implement targeted interventions to address potential risks before they lead to actual incidents. Encouraging a robust reporting culture is essential for capturing valuable information that contributes to continuous improvement in workplace safety.

16.3.3.5 *Safety Compliance Percentage*

The safety compliance percentage is a key performance indicator that measures the extent to which employees and operations comply with established safety procedures, regulations, and standards within an organization. This metric provides insight into the overall adherence to safety protocols. The Safety Compliance Percentage is calculated by comparing the number of observed compliance actions or behaviors with the total number of required or expected actions. The formula is as follows:

$$\text{Safety Compliance Percentage} = \left(\frac{\text{Number of Observed Compliance Actions}}{\text{Total Number of Required Actions}} \right) \times 100 \quad (16.5)$$

where:

- Number of Observed Compliance Actions: This refers to the actions or behaviors that are in compliance with established safety procedures. These actions are typically observed through inspections, audits, or other monitoring mechanisms.
- Total Number of Required Actions: This represents the total number of actions or behaviors that are expected or required according to safety protocols, regulations, or standards.
- 100: This factor is included to express the result as a percentage.

Interpreting the Safety Compliance Percentage: A higher safety compliance percentage indicates a stronger culture of safety compliance within the organization. It suggests that employees are consistently following established safety procedures and guidelines. A lower safety compliance percentage may signal areas of concern where compliance with safety protocols is lacking. It could prompt the organization to investigate the root causes and implement corrective actions.

Monitoring the safety compliance percentage allows organizations to identify trends, measure the effectiveness of safety training programs, and target interventions to improve compliance in specific areas. Regular assessments and adjustments to safety protocols may be necessary to ensure ongoing adherence and continuous improvement in workplace safety. Additionally, this KPI can be complemented by other safety metrics to provide a more comprehensive view of overall safety performance.

16.3.3.6 Emergency Response Time

Emergency response time is a crucial metric that measures the time it takes for an organization to respond to and address emergency situations. This KPI is particularly important in industries where rapid and effective emergency response is essential for minimizing the impact of incidents on safety, the environment, and property. The emergency response time is calculated as the time elapsed from the initiation of an emergency to the point at which the emergency response team is fully deployed and the situation is under control. The formula is straightforward:

$$Emergency\,Response Time = Time of\; Full Deployment - Time of\; Emergency\, Initiation \quad (16.6)$$

where:

- Time of Full Deployment: This is the point in time when the emergency response team is fully mobilized and actively addressing the emergency. It includes the time taken to assemble the team, arrive at the scene, and initiate response actions.
- Time of Emergency Initiation: This marks the beginning of the emergency event or incident, often signaled by the moment when the emergency is officially recognized or reported.

Interpreting Emergency Response Time: A shorter emergency response time is generally preferred, as it indicates a more efficient and effective emergency response. Quick response times can contribute to minimizing the impact of emergencies and reducing potential harm. A longer emergency response time may suggest areas for improvement in the organization's emergency preparedness and response procedures. It could prompt a review of response protocols, training, and resource allocation.

Factors Affecting Emergency Response Time:

- Training and Preparedness: The level of training and preparedness of the emergency response team members can significantly impact response times. Well-trained teams are more likely to respond quickly and effectively.
- Communication Systems: The efficiency of communication systems, including notification and alert systems, can influence how quickly an organization becomes aware of an emergency and initiates the response.
- Resource Availability: The availability and accessibility of emergency response resources, such as equipment and personnel, play a crucial role in response times.
- Emergency Response Plan: The organization's emergency response plan, including clearly defined roles, responsibilities, and procedures, is instrumental in facilitating a swift and coordinated response.

Regularly monitoring and analyzing emergency response time helps organizations identify opportunities for improvement in their emergency response capabilities. It is essential to conduct periodic drills, training sessions, and reviews of response protocols to enhance overall emergency preparedness and reduce response times.

16.3.3.7 Safety Training Completion Rate

The safety training completion rate is a key performance indicator that measures the percentage of employees who have successfully completed required safety training within a specified time frame. This metric reflects the organization's commitment to providing adequate safety education and ensuring that employees are well-informed about safety procedures and protocols. The safety training completion rate is calculated using the following formula:

$$\text{Safety Training Completion Rate} = \left(\frac{\text{Number of Employees Completing Safety Training}}{\text{Total Number of Employees Requiring Training}} \right) \times 100 \quad (16.7)$$

where:

- Number of Employees Completing Safety Training: This represents the total number of employees who have successfully completed the required safety training within the specified time frame.
- Total Number of Employees Requiring Training: This is the total number of employees who are required to undergo the specified safety training.
- 100: This factor is included to express the result as a percentage.

Interpreting the Safety Training Completion Rate: A higher safety training completion rate indicates a strong commitment to safety training within the organization. It suggests that a significant proportion of employees have received the necessary training to enhance their awareness and knowledge of safety procedures. A lower safety training completion rate may indicate challenges in delivering or tracking safety training programs. It could prompt the organization to investigate the reasons for incomplete training and implement strategies to improve completion rates.

Factors Affecting Safety Training Completion Rate:

- Training Accessibility: The ease of access to training materials, resources, and delivery methods can influence completion rates. Providing accessible and user-friendly training options enhances participation.
- Training Relevance: Ensuring that safety training programs are relevant to employees' roles and responsibilities increases engagement and motivation to complete the training.
- Communication: Clear communication about the importance of safety training, deadlines, and consequences for non-compliance can encourage employees to prioritize and complete required training.
- Training Delivery Methods: Utilizing a variety of training methods, including online modules, in-person sessions, and

hands-on training, accommodates different learning preferences and schedules.

- Management Support: Leadership and management support for safety initiatives, including training, reinforces the importance of safety within the organization and encourages employee participation.

Regularly monitoring the safety training completion rate allows organizations to assess the effectiveness of their safety training programs, identify any gaps or challenges, and implement improvements to enhance overall safety awareness and compliance. It is an essential component of a proactive safety culture within the workplace.

16.3.3.8 Workplace Inspection Finding Closure Rate

The workplace inspection finding closure rate is a key performance indicator that measures the efficiency of an organization in addressing and closing out safety-related findings identified during workplace inspections. This metric is crucial for ensuring that identified hazards and safety issues are promptly resolved to maintain a safe working environment. The workplace inspection finding closure rate is calculated using the following formula:

$$\text{Workplace Inspection Findings Closure Rate} = \left(\frac{\text{Number of Closed Findings}}{\text{Total Number of Findings Identified}} \right) \times 100 \quad (16.8)$$

where:

- Number of Closed Findings: This represents the total number of safety findings or issues that have been successfully addressed, corrected, and closed out following a workplace inspection.
- Total Number of Findings Identified: This is the overall number of safety findings or issues identified during workplace inspections, including those that have been closed and those that remain open.
- 100: This factor is included to express the result as a percentage.

Interpreting the Workplace Inspection Finding Closure Rate: A higher closure rate indicates an effective and proactive approach to addressing safety issues identified during inspections. It suggests that the organization is promptly taking corrective actions to resolve hazards and improve safety conditions. A lower closure rate may signal challenges in the organization's ability to address and close out identified safety issues. It could prompt a review of the processes for addressing findings and the allocation of resources to improve closure rates.

Factors Affecting Workplace Inspection Finding Closure Rate:

- Resource Allocation: The availability of resources, including personnel, tools, and materials, can impact the organization's ability to address and close out inspection findings in a timely manner.
- Communication: Effective communication between those responsible for addressing findings and the inspection team is essential. Clear communication channels help ensure that corrective actions are understood and implemented correctly.
- Priority Setting: The organization's prioritization of safety issues can influence which findings are addressed first. High-priority findings that pose immediate risks may be addressed more urgently.
- Follow-Up Procedures: Having established procedures for follow-up on inspection findings, including tracking and reporting mechanisms, contributes to a more systematic and organized approach to closure.
- Training and Awareness: Ensuring that employees are aware of the importance of addressing inspection findings and providing necessary training on corrective actions can improve closure rates.

Regularly monitoring the workplace inspection finding closure rate allows organizations to evaluate the efficiency of their safety management system, identify areas for enhancement, and implement strategies to improve closure rates and overall safety performance. It is a key element of a continuous improvement approach to workplace safety.

16.3.3.9 *Safety Culture Surveys*

Safety culture surveys are tools used to assess and measure the safety culture within an organization. Safety culture refers to the shared values, beliefs, attitudes, and behaviors regarding safety that exist among employees at all levels. Conducting safety culture surveys helps organizations understand the perceptions and attitudes of employees toward safety, identify strengths and weaknesses in the safety culture, and guide efforts to enhance overall safety performance. While there is not a specific formula for safety culture surveys, the process generally involves the following steps:

- Survey Design: Develop a survey instrument that includes questions addressing various aspects of safety culture, such as leadership commitment to safety, communication, employee involvement, perception of risk, and overall safety awareness.
- Anonymity and Confidentiality: To promote honest and open replies, make sure the survey is performed securely. This helps in obtaining accurate feedback without fear of reprisal.
- Distribution: Distribute the survey to all levels of employees within the organization. Use various channels, such as online surveys, paper-based surveys, or a combination, to reach a broad audience.
- Data Collection: Collect and compile survey responses. Use survey software or other tools to organize and analyze the data efficiently.
- Analysis: Analyze survey results to identify patterns, trends, and areas of strength or concern. Look for common themes in the responses to gain insights into the organization's safety culture.
- Benchmarking: Compare survey results against industry benchmarks or established norms to assess how the organization's safety culture compares to recognized standards.
- Action Planning: Develop action plans based on survey findings. Identify areas that require improvement and establish specific goals, strategies, and timelines for enhancing the safety culture.
- Communication of Results: Share survey results with employees and stakeholders. Transparency in sharing findings helps

build trust and demonstrates the organization's commitment to addressing safety concerns.
- Implementation of Changes: Implement the action plans developed in response to survey findings. This may involve changes in policies, procedures, training programs, or communication strategies.
- Follow-Up Surveys: Conduct follow-up surveys at regular intervals to track changes in safety culture over time. This allows organizations to measure the effectiveness of implemented changes and identify new areas for improvement.

Safety culture surveys provide a valuable tool for organizations to assess, understand, and improve their safety culture. Regular surveys help to build a proactive safety atmosphere that prioritizes employee well-being and fosters a more secure workplace. Key aspects of safety culture surveys as the following:

- Leadership Commitment: Assess the extent to which leadership demonstrates a commitment to safety and the importance of employee well-being.
- Communication: Evaluate the effectiveness of communication channels related to safety information and the clarity of safety messages.
- Employee Involvement: Measure the level of employee involvement in safety initiatives and decision-making processes.
- Perception of Risk: Understand how employees perceive and assess risks within the organization.
- Continuous Improvement: Gauge the organization's commitment to continuous improvement in safety practices.

16.3.3.10 Safety Meeting Attendance

Safety meeting attendance is a key performance indicator that measures the level of employee participation in safety meetings or briefings. Regular safety meetings are a fundamental component of a proactive safety program, providing a platform for communication, training, and the dissemination of important safety information. Monitoring safety meeting attendance helps organizations ensure that employees are engaged and informed about safety procedures and initiatives. The formula for calculating safety meeting attendance is straightforward:

$$Safety\ Meetings\ Attendance = \left(\frac{Number\ of\ Attendees}{Total\ Number\ of\ Employees}\right) \times 100 \quad (16.9)$$

where:

- Number of Attendees: This represents the total number of employees who attended a safety meeting or briefing.
- Total Number of Employees: This is the overall number of employees in the organization, including those who were eligible to attend the safety meeting.
- 100: This factor is included to express the result as a percentage.

Interpreting Safety Meeting Attendance: A higher safety meeting attendance percentage indicates strong employee engagement in safety initiatives. It suggests that a significant portion of the workforce actively participates in safety meetings. A lower safety meeting attendance percentage may indicate challenges in engaging employees in safety-related discussions. It could prompt the organization to assess the effectiveness of communication strategies or identify barriers to attendance.

Factors affecting Safety Meeting Attendance:

- Communication: Effective communication about the importance of safety meetings and the relevance of the information discussed can enhance attendance.
- Meeting Scheduling: Choosing appropriate times and frequencies for safety meetings that accommodate employees' schedules contributes to higher attendance rates.
- Meeting Content: Ensuring that safety meetings are engaging, informative, and directly relevant to employees' roles and responsibilities encourages attendance.
- Incentives and Recognition: Providing incentives or recognition for regular attendance can motivate employees to participate in safety meetings.
- Leadership Involvement: Demonstrating leadership support for safety initiatives by actively participating in or leading safety meetings can positively influence attendance.

- Variety in Meeting Formats: Incorporating different formats, such as interactive sessions, guest speakers, or hands-on activities, can make safety meetings more engaging and encourage attendance.

Regularly monitoring safety meeting attendance allows organizations to evaluate the efficiency of their safety communication strategies, identify areas for enhancement, and ensure that critical safety information is reaching the entire workforce. It is an essential aspect of fostering a safety-conscious culture within the organization.

16.3.3.11 Equipment Failure Rate

The equipment failure rate is a key performance indicator that measures the frequency at which equipment within an organization fails or experiences malfunctions over a specific period. This metric is crucial for assessing the reliability of equipment and identifying areas for improvement in maintenance and reliability programs. The equipment failure rate is calculated using the following formula:

$$Equipment\ Failure\ Rate = \left(\frac{Number\ of\ Equipment\ Failures}{Total\ Operating\ Time\ of\ Equipment}\right) \times 100 \quad (16.10)$$

where:

- Number of Equipment Failures: This represents the total count of equipment failures or malfunctions that occurred during the specified time frame.
- Total Operating Time of Equipment: This is the cumulative time that the equipment was in operation during the same time frame. It can be expressed in hours, days, or another relevant unit.
- 100: This factor is included to express the result as a percentage.

Interpreting the Equipment Failure Rate: A lower equipment failure rate indicates better equipment reliability, suggesting that the organization's maintenance and reliability programs are effective in preventing or minimizing equipment failures. A higher equipment failure rate suggests that the organization may be experiencing challenges in maintaining the reliability of its equipment. It signals a need for closer scrutiny of maintenance practices and potential improvements.

Factors Affecting Equipment Failure Rate:

- Proactive Maintenance Practices: Implementing proactive maintenance strategies, such as preventive and predictive maintenance, can contribute to reducing the frequency of equipment failures.
- Quality of Equipment: The design, manufacturing quality, and age of equipment can influence its reliability. Well-maintained and newer equipment may have lower failure rates.
- Training and Skill of Maintenance Personnel: Adequately trained and skilled maintenance personnel are better equipped to identify and address potential issues before they lead to equipment failures.
- Monitoring and Condition-Based Maintenance: Regular monitoring of equipment conditions and the use of condition-based maintenance approaches help in identifying and addressing issues before they escalate.
- Equipment Redundancy: Having redundancy or backup systems in place can mitigate the impact of equipment failures and contribute to overall system reliability.
- Root Cause Analysis: Conducting thorough root cause analyses for equipment failures helps in identifying the underlying reasons and implementing corrective actions to prevent recurrence.
- Reliability Centered Maintenance (RCM): Implementing RCM methodologies involves analyzing the functions and potential failure modes of equipment to optimize maintenance strategies.
- Equipment Lifecycle Management: Properly managing the lifecycle of equipment, including retirement and replacement planning, helps maintain an optimal and reliable equipment fleet.

Regularly monitoring the equipment failure rate and analyzing trends over time allows organizations to optimize maintenance strategies, make data-driven decisions, and enhance overall equipment reliability. This KPI is especially critical in industries where equipment performance directly impacts operational efficiency, safety, and costs.

16.3.3.12 Hazard Identification and Resolution Rate

The hazard identification and resolution rate is a key performance indicator that measures an organization's effectiveness in identifying potential hazards and resolving them to prevent incidents and promote a safe working environment. This KPI is crucial for proactive risk management and ensuring continuous improvement in safety practices. The hazard identification and resolution rate is calculated using the following formula:

$$\textit{Hazard Identification and Resolution Rate} = \left(\frac{\textit{Number of Resolved Hazards}}{\textit{Total Number of Identified Hazards}} \right) \times 100 \quad (16.11)$$

where:

- Number of Resolved Hazards: This represents the total count of identified hazards for which corrective actions have been taken to mitigate the risk and resolve the hazard.
- Total Number of Identified Hazards: This is the cumulative number of hazards that have been identified during a specific period, which may be based on inspections, risk assessments, incident investigations, or other safety processes.
- 100: This factor is included to express the result as a percentage.

Interpreting the Hazard Identification and Resolution Rate: A higher hazard identification and resolution rate indicates that the organization is effectively identifying potential hazards and taking corrective actions to resolve them. It reflects a proactive approach to risk management. A lower hazard identification and resolution rate may suggest that there are challenges in promptly addressing identified hazards. It could prompt a review of hazard identification processes and the effectiveness of corrective actions.

Factors affecting Hazard Identification and Resolution Rate:

- Effective Reporting Systems: Establishing and promoting a robust reporting system for hazards encourages employees to report potential risks, contributing to a higher identification rate.
- Regular Inspections and Audits: Conducting regular workplace inspections and safety audits helps identify hazards before they escalate and ensures that corrective actions are taken promptly.
- Employee Training: Providing training to employees on haz ard identification, reporting procedures, and the importance of proactive risk management enhances the overall hazard identification process.
- Root Cause Analysis: Conducting thorough root cause analyses for identified hazards helps in understanding the underlying reasons and implementing effective corrective actions.
- Management Commitment: Demonstrating leadership commitment to safety and providing the necessary resources for hazard resolution fosters a safety culture that prioritizes risk mitigation.
- Continuous Improvement: Establishing a culture of continuous improvement encourages ongoing efforts to enhance hazard identification processes and the effectiveness of corrective actions.

Regularly monitoring the hazard identification and resolution rate allows organizations to track their progress in creating a safe work environment, identify trends, and implement targeted interventions to improve safety practices. This KPI aligns with the principles of a proactive safety culture and supports the organization's commitment to the well-being of its employees.

16.3.3.13 Safety Budget Compliance

Safety budget compliance is a key performance indicator that measures the extent to which an organization adheres to its allocated budget for safety-related activities and initiatives. It reflects the financial discipline and commitment to safety management within the organization. The safety budget compliance is calculated using the following formula:

$$Safety\ Budget\ Compliance = \left(\frac{Actual\ Safety\ Expenditures}{Planned\ Safety\ Budget}\right) \times 100 \quad (16.12)$$

where:

- Actual Safety Expenditures: This represents the total amount of money spent on safety-related activities, programs, training, equipment, and other safety initiatives during a specific period.
- Planned Safety Budget: This is the budget allocated or planned for safety-related expenditures within the same time frame. It represents the financial resources designated for safety initiatives.
- 100: This factor is included to express the result as a percentage.

Interpreting Safety Budget Compliance: A higher safety budget compliance percentage indicates that the organization is effectively managing its safety budget and adhering to the planned financial allocations for safety initiatives. A lower safety budget compliance percentage may suggest challenges in financial planning, resource allocation, or control over safety-related expenditures. It could prompt a review of budgeting processes and financial management practices.

Factors Affecting Safety Budget Compliance:

- Accurate Budgeting: Ensuring that safety budgets are accurately calculated and aligned with the organization's safety objectives contributes to higher compliance rates.
- Resource Allocation: Properly allocating financial resources to safety programs, training, equipment, and other initiatives ensures that adequate funding is available to support safety measures.
- Cost Control Measures: Implementing cost control measures and monitoring expenses help prevent budget overruns and ensure that safety expenditures align with the planned budget.

- Return on Investment (ROI): Assessing the ROI of safety expenditures helps organizations determine the effectiveness of safety programs and justifies budget allocations.
- Prioritization of Safety: A strong commitment to safety at all levels of the organization reinforces the importance of adhering to safety budgets and ensures that safety remains a priority.
- Regular Financial Reviews: Conducting regular financial reviews and reconciliations helps identify discrepancies, track expenses, and maintain financial accountability.

Monitoring safety budget compliance is essential for demonstrating financial responsibility and ensuring that the organization has the necessary resources to support a robust safety program. It aligns financial management practices with the organization's commitment to creating a safe and healthy work environment. Regular reviews of safety expenditures and budget compliance contribute to the continuous improvement of safety initiatives and resource allocation strategies.

16.3.3.14 Contractor Safety Performance

Contractor safety performance is a key performance indicator that measures the safety performance of external contractors working within an organization. It assesses how well contractors adhere to safety standards, regulations, and practices while carrying out work on behalf of the organization. Monitoring contractor safety performance is critical for ensuring a safe working environment and preventing incidents that could impact both the contractor's workforce and the organization's employees. It is essential for fostering a culture of safety throughout the entire workforce, including external contractors who contribute to the organization's operations. The assessment of contractor safety performance involves several factors:

- Incident Rates: Track the number of safety incidents, injuries, and near misses involving contractors. Compare these rates to established benchmarks or the organization's own safety performance standards.
- Compliance With Safety Policies: Evaluate the extent to which contractors comply with the organization's safety policies, procedures, and guidelines. This includes adherence to safety regulations and industry standards.

- Safety Training and Certification: Assess the level of safety training and certification that contractors provide to their employees. Check that their team has the essential skills and expertise to do the job safely.
- Safety Audits and Inspections: Audit and inspect contractor work locations on a regular basis for safety to identify potential hazards, assess compliance, and ensure that safety protocols are being followed.
- Emergency Response Preparedness: Evaluate the contractor's emergency response preparedness, including their ability to respond to incidents promptly and effectively.
- Documentation and Reporting: Review the contractor's documentation and reporting processes for safety-related incidents and activities. Ensure that incidents are reported accurately and in a timely manner.
- Safety Culture: Assess the safety culture within the contractor's organization. A positive safety culture contributes to a proactive approach to safety and a commitment to continuous improvement.
- Communication and Collaboration: Evaluate the effectiveness of communication and collaboration between the organization and its contractors regarding safety expectations, updates, and changes in safety protocols.
- Performance Reviews and Scorecards: Implement performance reviews and scorecards for contractors based on safety performance metrics. Provide feedback and address any areas for improvement.
- Contractor Selection Criteria: Include safety performance as a key criterion when selecting contractors. Prioritize contractors with a strong safety track record.
- Continuous Improvement Initiatives: Encourage contractors to engage in continuous improvement initiatives related to safety. Foster a collaborative approach to address safety challenges and enhance performance.

16.4 Advantages of Safety Key Performance Indicators

Safety key performance indicators offer several advantages for organizations committed to maintaining a safe and healthy work environment. In summary, safety KPIs play a crucial role in promoting a proactive safety culture, driving continuous improvement, and ensuring the well-being of employees. They provide a structured approach to managing safety performance and contribute to the overall success and sustainability of the organization. Some key advantages are as follows:

- Performance Measurement: Safety KPIs provide quantifiable metrics that allow organizations to measure their safety performance objectively. This enables the organization to track progress over time and set realistic safety goals.
- Goal Setting and Benchmarking: Safety KPIs assist in setting specific, measurable, achievable, relevant, and time-bound (SMART) safety goals. Organizations can benchmark their performance against industry standards or their own historical data.
- Continuous Improvement: Organizations may determine areas for enhancement and conduct focused actions by frequently monitoring safety KPIs. This contributes to a culture of constant improvement in safety practices.
- Risk Identification and Mitigation: Safety KPIs help identify potential risks and hazards in the workplace. Addressing these risks proactively can prevent incidents and reduce the likelihood of injuries, illnesses, or accidents.
- Decision-Making Support: Safety KPIs provide management with data-driven insights for informed decision-making. This allows leaders to allocate resources effectively, prioritize safety initiatives, and address critical safety concerns.
- Employee Engagement: Transparent communication of safety KPIs fosters a sense of responsibility and engagement among employees. It helps them understand the importance of safety and encourages active participation in safety initiatives.
- Compliance Monitoring: Safety KPIs assist organizations in monitoring compliance with safety regulations, standards, and internal policies. This ensures that the organization meets its legal obligations and operates within established safety guidelines.
- Demonstrating Accountability: Safety KPIs create a framework for holding individuals, teams, and the organization as a whole accountable for safety performance. This

accountability reinforces a commitment to safety from top leadership to frontline workers.

- Resource Allocation Efficiency: By identifying areas with higher safety risks, organizations can allocate resources more efficiently. This includes directing resources to training programs, safety equipment, and other initiatives that address specific safety challenges.
- Enhanced Communication: Safety KPIs facilitate communication between different levels of the organization. Regular reporting on safety metrics promotes transparency and ensures that everyone is aware of safety priorities and progress.
- Cost Savings: A strong safety performance, as indicated by KPIs, can result in cost savings. Fewer accidents and injuries mean reduced medical costs, lower workers' compensation claims, and improved productivity.
- Reputation Management: Demonstrating a commitment to safety through positive KPIs enhances an organization's reputation. This can be important for attracting and retaining employees, winning contracts, and building trust with stakeholders.
- Legal Protection: Maintaining strong safety KPIs provides a degree of legal protection. In the event of an incident, a documented history of compliance and safety performance can support the organization's defense.
- Sustainability and Corporate Social Responsibility: Safety KPIs contribute to an organization's sustainability efforts and corporate social responsibility initiatives. Demonstrating a commitment to the well-being of employees aligns with broader societal expectations.
- Overall Organizational Performance: A safe workplace contributes to overall organizational performance. Healthy and motivated employees are more likely to be productive, reducing absenteeism and turnover, and positively impacting the organization's bottom line.

16.5 Disadvantages of Safety Key Performance Indicators

While safety KPIs offer valuable insights into an organization's safety performance, there are also potential disadvantages and challenges associated with their use. It is important to be aware of these considerations to implement safety KPIs effectively. To mitigate these disadvantages, organizations should approach the use of safety KPIs with a balanced perspective, considering both leading and lagging indicators, emphasizing a strong safety culture, and continually refining their approach based on feedback and evolving organizational needs. Regular reviews of safety metrics and open communication can contribute to a more effective use of safety KPIs. Some of the disadvantages are as follows:

- Overemphasis on Metrics: Focusing too much on specific safety metrics may lead to a situation where meeting targets becomes the primary goal, potentially overshadowing the broader safety culture and individual behaviors.
- Metric Manipulation: There may be a risk of employees or management manipulating safety data to meet KPI targets. This could include underreporting incidents or misclassifying incidents to present a more favorable safety performance.
- Unintended Consequences: Setting KPIs without considering potential unintended consequences may lead to behaviors that prioritize meeting KPIs over addressing the root causes of safety issues. This can result in a superficial improvement in safety performance.
- Focus on Lagging Indicators: Some safety KPIs, such as injury rates, are lagging indicators that only reflect past performance. Relying solely on lagging indicators may hinder proactive identification and mitigation of emerging risks.
- Lack of Context: Safety KPIs, when considered in isolation, may lack the context needed to understand the underlying factors contributing to safety performance. It is important to complement KPIs with qualitative information and insights.
- Benchmarking Challenges: Benchmarking safety KPIs against industry standards can be challenging due to variations in reporting practices, definitions, and industry-specific factors. This makes accurate comparisons difficult.
- Subjectivity in Reporting: The subjectivity in reporting incidents and near misses can introduce variability in safety KPIs. Different individuals or teams may interpret and report incidents differently, affecting the accuracy of the data.
- Complexity of Safety Performance: Safety is a complex and multifaceted aspect of organizational performance. Relying solely on a limited set of safety KPIs may oversimplify the

understanding of safety issues and hinder comprehensive risk management.

- Potential for Neglecting Leading Indicators: Placing too much emphasis on lagging indicators may result in neglecting leading indicators that provide insights into proactive measures and the effectiveness of preventive actions.
- Data Quality and Accuracy: The accuracy and reliability of safety data depend on the quality of reporting systems. Inaccurate or incomplete data can lead to misleading safety KPIs and compromise decision-making.
- Time Lags in Reporting: Reporting delays can impact the timeliness of safety KPIs, especially if there are delays in recording incidents or if there are reporting lags from remote or decentralized work locations.
- Resistance to Reporting: If there is a perception that reporting incidents or near misses may have negative consequences, employees may be hesitant to report accurately, leading to an underestimation of safety risks.
- Static Nature of KPIs: Safety KPIs, once established, may become static and may not evolve with changes in organizational structure, processes, or industry dynamics. Regular review and adjustment are necessary for relevance.
- Limited Scope of Traditional KPIs: Traditional safety KPIs may not capture emerging risks associated with new technologies, changes in work processes, or evolving workplace conditions, limiting their ability to address contemporary safety challenges.

16.6 Application of Safety Key Performance Indicators

Safety key performance indicators (KPIs) find applications across various industries and organizational settings to monitor, assess, and improve safety performance [7]. The applications of safety KPIs are diverse and encompass various aspects of safety management. They play a crucial role in fostering a proactive safety culture and ensuring the well-being of employees [8]. Some common applications of safety KPIs are as follows:

- Performance Monitoring: Safety KPIs are used to monitor the overall safety performance of an organization. Metrics such as injury rates, incident rates, and near-miss reports provide a quantitative assessment of safety performance over time.
- Benchmarking: Organizations use safety KPIs to benchmark their safety performance against industry standards, best practices, or their own historical data. This allows for comparisons that can inform improvement strategies.
- Compliance Management: Safety KPIs help organizations ensure compliance with safety regulations, standards, and internal policies. Monitoring compliance metrics provides insights into adherence to safety protocols and legal requirements.
- Goal Setting and Target Achievement: Safety KPIs are instrumental in setting specific safety goals and targets. By establishing measurable objectives, organizations can work toward achieving and surpassing these targets, contributing to a proactive safety culture.
- Risk Identification and Mitigation: Leading safety indicators, such as the number of safety inspections conducted or the completion of safety training programs, help identify potential risks before they escalate. This proactive approach allows for timely risk mitigation.
- Resource Allocation: Safety KPIs aid in allocating resources efficiently. Organizations can identify high-risk areas and allocate resources such as training, equipment, and personnel to address specific safety challenges.
- Decision-Making Support: Data from safety KPIs supports informed decision-making. Leaders can use this information to allocate budgets, prioritize safety initiatives, and make strategic decisions to improve overall safety performance.
- Employee Engagement: Safety KPIs contribute to employee engagement by fostering a sense of responsibility and accountability for safety. Transparent communication about safety metrics encourages employees to actively participate in safety initiatives.
- Continuous Improvement Initiatives: Safety KPIs drive continuous improvement by highlighting areas for enhancement. Organizations can use the data to identify trends,

analyze root causes, and implement targeted interventions to improve safety practices.

- Performance Reviews and Incentives: Safety KPIs are often used in performance reviews for individuals, teams, or departments. Recognizing and rewarding positive safety performance encourages a culture of safety and reinforces good safety practices.
- Communication and Reporting: Safety KPIs support effective communication about safety performance within the organization. Regular reporting on safety metrics ensures that relevant stakeholders are informed about safety priorities and progress.
- Training Effectiveness: KPIs related to training completion rates and assessments help assess the effectiveness of safety training programs. This information guides adjustments and improvements in training content and delivery.
- Contractor Management: Organizations use safety KPIs to assess the safety performance of contractors and subcontractors. This helps in selecting and managing external partners with strong safety records.
- Emergency Response Planning: Metrics related to emergency response times and preparedness contribute to assessing and improving the effectiveness of emergency response plans.
- Reputation Management: Positive safety KPIs contribute to an organization's reputation. Demonstrating a commitment to safety is important for attracting and retaining talent, winning contracts, and building trust with stakeholders.

16.7 How to Make Safety Key Performance Indicators More Efficient?

Making safety KPIs more efficient involves optimizing their design, implementation, and utilization to drive meaningful improvements in safety performance. Organizations can enhance the efficiency and effectiveness of their safety KPIs, leading to improved safety outcomes and a stronger safety culture within the workplace. Here are several strategies to enhance the efficiency of safety KPIs:

- Align With Organizational Goals: Ensure that safety KPIs align with overall organizational goals and objectives. This alignment reinforces the importance of safety within the broader organizational context and encourages a unified approach.
- Select Relevant and Meaningful KPIs: Choose KPIs that are relevant to the organization's specific safety risks and priorities. Focus on indicators that provide actionable insights and contribute to a comprehensive understanding of safety performance.
- Use Leading and Lagging Indicators: To offer a comprehensive assessment of safety performance, balance the application of leading indicators with lagging indicators. Leading indicators help in proactive risk management.
- Involve Employees in KPI Selection: Engage employees in the selection of safety KPIs. Involving frontline workers ensures that KPIs reflect their experiences, concerns, and insights, promoting a sense of ownership and engagement.
- Set Realistic and Achievable Targets: Establish realistic and achievable targets for safety KPIs. Unrealistic goals can lead to frustration and demotivation. Targets should challenge the organization to improve while being attainable with concerted effort.
- Implement Continuous Monitoring: Continuously monitor safety KPIs rather than relying solely on periodic reviews. Real-time or near-real-time monitoring enables prompt identification of trends and issues, facilitating timely interventions.
- Utilize Technology: Leverage technology, such as data analytics tools and safety management software, to systematize data collection, analysis, and reporting. Technology streamlines the monitoring process and enhances accuracy.
- Provide Regular Feedback: Establish a system for providing regular feedback on safety KPIs to relevant stakeholders. Transparent communication fosters accountability, encourages improvement, and reinforces a safety-focused culture.
- Encourage Near-Miss Reporting: Include KPIs for reporting near-misses. Encouragement of near-miss reporting gives significant insights into possible dangers and enables proactive risk control.

- Integrate Safety KPIs Into Performance Reviews: Include safety KPIs as part of performance reviews for individuals, teams, and departments. Linking safety performance-to-performance evaluations reinforces the importance of safety.
- Promote Positive Reinforcement: Recognize and celebrate achievements and improvements in safety performance. Positive reinforcement, such as safety awards or recognition programs, motivates employees and teams to maintain and exceed safety standards.
- Regularly Review and Update KPIs: Evaluate and revise safety KPIs on a regular basis so that they stay relevant and connected with organizational aims. As the organization evolves, so too should its safety performance metrics.
- Provide Training and Support: Ensure that employees understand the importance of safety KPIs and receive adequate training to use safety tools and systems effectively. This empowers employees to actively contribute to safety goals.
- Conduct Root Cause Analyses: When safety KPIs indicate areas for improvement, conduct thorough root cause analyses to identify underlying issues. Addressing root causes is essential for sustained improvements.
- Encourage a Reporting Culture: Foster a culture that encourages open reporting of safety incidents, concerns, and observations. A reporting culture promotes transparency and ensures that safety KPIs reflect the true state of safety performance.

References

1. Louvar, J.L., Guidance for safety performance indicators. *Process Saf. Prog.*, 29, 4, 387–388, 2010.
2. Demareva, E.E., Key performance indicators of industrial safety of enterprises. *Наука Красноярья*, 11, 3, 7–16, 2022.
3. Mohammadfam, I., Kamalinia, M., Momeni, M., Golmohammadi, R., Hamidi, Y., Soltanian, A., Evaluation of the quality of occupational health and safety management systems based on key performance indicators in certified organizations. *Saf. Health Work*, 8, 2, 156–161, 2017.
4. Tugnoli, A., Landucci, G., Cozzani, V., Quantitative inherent safety assessment by key performance indicators (KPIs). *Chem. Eng. Trans.*, 17, 457–462, 2009.

5. Fuller, C.W., Key performance indicators for benchmarking health and safety management in intra-and inter-company comparisons. *Benchmarking Qual. Manage. Technol.*, 4, 3, 165–174, 1997.
6. Gerbec, M. and Kontić, B., Safety related key performance indicators for securing long-term business development–A case study. *Saf. Sci.*, 98, 77–88, 2017.
7. Shaikh, A.Y., Osei-Kyei, R., Hardie, M., A critical analysis of safety performance indicators in construction. *Int. J. Build. Pathol. Adapt.*, 39, 3, 547–580, 2021.
8. Mahmoud, A.S., Ahmad, M.H., Yatim, Y.M., Dodo, Y.A., Key performance indicators (KPIS) to promote building developers safety performance in the construction industry. *J. Ind. Eng. Manage. (JIEM)*, 13, 2, 371–401, 2020.

17

Occupational Health, Safety Management Systems, and Working Conditions

Abstract
This chapter explains the fundamental ideas required for effective use of occupational health and safety management systems (OHSMSs). It also highlights the key components of OHSMS, their merits and demerits, and their applicability in different places. In the latter part, ISO 45001:2018 standards have been discussed along with their applicability, pros and cons. Lastly, the Occupational Safety, Health, and Working Conditions (OSHWC) Code was discussed. The rights of an employee as well as employers under this code have been briefly discussed, along with their applicability.

Keywords: Occupational health and safety, ISO 45001:2018, OSHWC code

17.1 Background

Ensuring the basic right to a healthy and secure place of employment is critical for preventing workplace incidents and diseases, as well as protecting and promoting workers' health and well-being. The basic concepts of the process of providing a safe and healthy workplace are the detection of hazards occurring in or from the work environment, as well as the evaluation and control of the related risks that might affect the well-being and health of workers. The increasingly complicated and fast-changing nature of the workplace in recent decades has necessitated a systems-based approach to managing and sustaining a healthy and secure place to work. For numerous decades, occupational health and safety management systems (OHSMSs) have played a critical role in effectively improving workplace occupational health and safety implementation by guaranteeing integration into corporate planning and development processes [1].

Abhishek Barua, Kanchan Kumari, Sumit Saha, Siddharth Jeet and Swastik Pradhan. Institutional and Industrial Safety Engineering Practices, (401–426) © 2025 Scrivener Publishing LLC

17.2 Introduction to Occupational Health and Safety Management Systems

An OHSMS is a collection of interconnected or interacting elements used to define and accomplish occupational health and safety (OH&S) goals and regulations [2]. The use of a systemic approach to OH&S management in the workplace guarantees that the degree of prevention and protection is constantly reviewed and maintained via suitable and timely changes (Figure 17.1).

Employers are advised to build adequate plans for the implementation of an OHSMS with the goal of establishing and sustaining a safe and healthy working environment, as well as fulfilling the OH&S obligations enforced by national laws and regulations [3].

17.2.1 Components of OHSMS

OHSMSs are systematic techniques that organizations use to manage and enhance safety and health at work. By implementing and continuously improving an OHSMS, organizations can create a safer and improved work environment, reduce the risk of occupational injuries and illnesses, and enhance overall workplace well-being. Certification standards such as ISO 45001 provide a framework for organizations to develop and implement

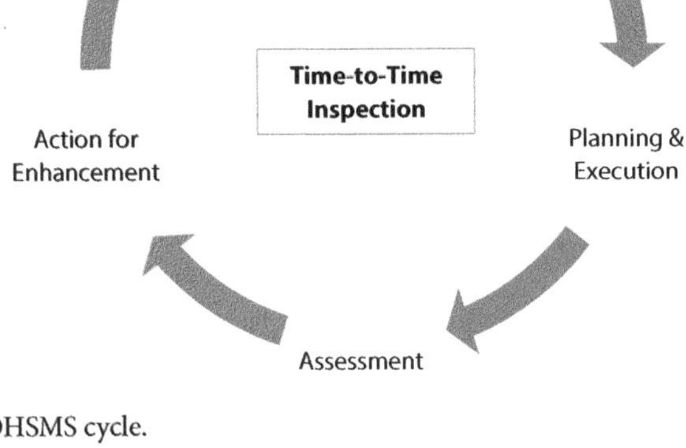

Figure 17.1 OHSMS cycle.

active occupational safety and health management systems [4]. The working of an OHSMS typically involves several key steps and components:

1. Policy Development: The process frequently starts with the creation of an OH&S policy. This policy establishes the framework for the OHSMS by outlining the organization's dedication to delivering a safe and healthy workplace.
2. Leadership and Commitment: Leadership commitment is crucial for the success of an OHSMS. Top management demonstrates a commitment to OH&S by integrating it into the overall organizational strategy, providing resources, and promoting a safety culture.
3. Hazard Identification and Risk Assessment: To detect possible risks in the workplace, organizations perform extensive hazard identification and risk assessments. This entails determining the probability and severity of risks connected with specific tasks, activities, and processes.
4. Legal Compliance: OHSMS guarantees that organizations follow relevant occupational safety and health regulations, laws, and industry standards. This includes regular monitoring of changes in legislation and updating policies and procedures accordingly.
5. Roles and Responsibilities: Clearly defined roles and responsibilities for OHS are established throughout the organization. This includes the assignment of responsibilities to individuals at various levels to ensure accountability for OH&S performance.
6. Training and Competence: OHSMS includes provisions for the training and competence of employees. This makes sure people have the skills and information needed to execute their jobs in a secure manner to occupational health and safety–related challenges.
7. Communication: Effective communication is essential in an OHSMS. This involves communicating the organization's OH&S policy, objectives, and procedures to all levels of the workforce. It also encourages open communication channels for reporting safety concerns and incidents.
8. Emergency Preparedness and Response: OHSMS includes plans and procedures for emergency preparedness and response. This ensures that employees are trained to respond

to emergencies, and there are measures in place to mitigate the impact of incidents.

9. Monitoring and Measurement: An OHSMS must include regular evaluation and tracking of OH&S performance. This involves the collection of data on incidents, near misses, and other relevant metrics to assess the efficiency of OH&S controls and processes.
10. Evaluation of Compliance: Ongoing evaluation ensures that the organization complies with appropriate OH&S laws and regulations. Regular audits and assessments help identify areas for improvement and corrective actions.
11. Documentation and Record-Keeping: OHSMS requires the creation and maintenance of documentation and records related to OH&S activities, procedures, policies, and performance. Proper documentation supports transparency and accountability.
12. Review and Improvement: OHSMS is a dynamic system that requires regular review and improvement. Top management conducts periodic reviews of the OHSMS to ensure its effectiveness and makes adjustments based on changing circumstances or new information.

17.2.2 Advantages of OHSMS

The application of an OHSMS in an industry provides several advantages, contributing to the safety of employees, the overall safety culture, and the success of the organization. Implementing an OHSMS in an industry offers a range of benefits that extend beyond compliance. It creates a safer, healthier workplace, positively influences the organization's reputation, and contributes to overall business success. Some key advantages are as follows:

1. Reduced Incidents and Injuries: An OHSMS helps identify and mitigate potential hazards and risks, leading to a reduction in workplace incidents and injuries. This contributes to a safer working environment for employees.
2. Improved Health and Well-being: The focus on occupational health within the management system promotes the well-being of employees, addressing factors that could impact their physical and mental health.
3. Legal Compliance: Implementing an OHSMS ensures that the organization is aware of and complies with relevant

OH&S laws and regulations. This helps avoid legal issues and penalties.

4. Enhanced Risk Management: An OHSMS's systematic methodology enables the detection, evaluation, and handling of OH&S concerns. Proactive risk management reduces the likelihood of incidents.
5. Cost Savings: An OHSMS can result in considerable cost savings by avoiding accidents and lowering the number of casualties. This includes lower workers' compensation claims, medical bills, and downtime costs.
6. Improved Productivity: A safer work environment tends to result in improved employee morale and job satisfaction. This, in turn, positively affects productivity, as motivated and healthy employees are generally more engaged and efficient.
7. Enhanced Reputation: Organizations with effective OHSMS are often perceived as socially responsible and committed to employee well-being. This positive reputation can enhance relationships with customers, suppliers, and other stakeholders.
8. Competitive Advantage: Certification or adherence to recognized OHSMS standards, such as ISO 45001, can provide a competitive advantage in the marketplace. It may be a requirement for certain contracts or partnerships.
9. Employee Engagement: Employee involvement in the OHSMS process develops a safety culture and encourages active engagement. Employees who are engaged are more inclined to make a contribution to a safer workplace.
10. Continuous Improvement: OHSMS frameworks emphasize a commitment to continuous improvement. Regular reviews, assessments, and updates to the system ensure that safety practices evolve with changing circumstances.
11. Leadership Commitment: Demonstrating leadership commitment to OH&S through the implementation of an OHSMS sets a positive example for employees and reinforces the importance of safety.
12. Improved Communication: The OHSMS promotes effective communication about safety policies, procedures, and performance within the organization. Clear communication is essential for creating a shared understanding of safety expectations.

13. Employee Training and Development: An OHSMS includes provisions for employee training on safety procedures and practices. Continuous learning and development contribute to a workforce that is knowledgeable about occupational health and safety.
14. Emergency Preparedness: Organizations are required under OHSMS frameworks to establish and put into effect plans for responding to emergencies. Employees are thus better equipped to respond to and handle situations.
15. Increased Employee Retention: Concentrating on OH&S can help increase satisfaction among workers and retention. Employees are more inclined to stay with a company that values their health.

17.2.3 Disadvantages of OHSMS

While implementing an OHSMS in an industry brings numerous benefits, there are also potential disadvantages and challenges that organizations may face. It is essential to recognize that the potential disadvantages of OHSMS can be mitigated with careful planning, effective communication, and a commitment to continuous improvement. A well-implemented OHSMS can significantly contribute to creating a safer and healthier work environment. Some disadvantages associated with OHSMS in industry are as follows:

1. Resource Intensive: Implementing and maintaining an OHSMS can be resource-intensive, requiring significant time, personnel, and financial investments.
2. Complexity of Implementation: Developing and implementing an OHSMS can be a complex process, especially for organizations with limited experience in management systems. The complexity may lead to challenges in understanding and meeting the requirements.
3. Perceived Bureaucracy: Some employees may perceive the documentation and procedural requirements of an OHSMS as bureaucratic and administrative, potentially leading to frustration and resistance.
4. Initial Costs: There are initial costs associated with implementing an OHSMS, including training, hiring consultants,

and developing documentation. These costs may be a barrier for some organizations, especially smaller ones.

5. Management Involvement: The commitment of top management is crucial for the success of an OHSMS. If leaders are not actively engaged and supportive, then the effectiveness of the system may be compromised.
6. Misalignment With Organizational Structure: The standardized nature of some OHSMS frameworks may not align perfectly with the specific organizational structure or industry practices, leading to challenges in implementation.
7. Documentation Overemphasis: In some cases, there may be a tendency to focus excessively on documentation to meet compliance requirements, potentially diverting attention from practical safety improvements.
8. Limited Flexibility: Strict adherence to the requirements of an OHSMS standard may limit the flexibility needed to address unique industry challenges or evolving safety risks.
9. Not a One-Size-Fits-All Solution: OHSMS standards provide a framework, but they may not address all the unique safety challenges of every industry or organization. Organizations must tailor their OHSMS to their specific needs.
10. Audit Pressure: The regular audits required for maintaining OHSMS certification may create additional pressure on employees, particularly if they perceive audits as punitive rather than as opportunities for improvement.
11. Focus on Compliance Rather Than Improvement: There is a risk that organizations may focus more on meeting certification requirements for compliance rather than on genuinely improving OH&S outcomes.
12. Overreliance on Documentation: Relying too heavily on documented procedures and processes may create a false sense of security. Effective safety management requires practical implementation beyond paperwork.
13. Employee Burnout: If the workload associated with implementing and maintaining an OHSMS is not well-managed, it can lead to employee burnout and decreased morale.
14. External Factors: External factors such as changes in regulations or industry standards may require organizations to continually update their OHSMS, adding complexity and potential disruption.

17.2.4 Application of OHSMS

OHSMS finds applications across various industries and organizations. Implementing an OHSMS has the primary purpose of creating a safe and healthy work environment, preventing work-related accidents and illnesses, and ensuring adherence to OH&S requirements [3]. Some key applications of OHSMS are as follows:

- Risk Management: OHSMS helps organizations identify, assess, and manage risks associated to occupational health and safety. By conducting risk assessments, organizations can proactively address potential hazards and implement control measures to minimize risks.
- Legal Compliance: Compliance helps organizations avoid legal penalties, fines, and reputational damage while promoting a culture of responsibility.
- Incident Prevention: OHSMS focuses on preventing incidents, accidents, and occupational diseases. Organizations can minimize the incidence of occupational injuries and illnesses by introducing preventive strategies and tackling fundamental causes.
- Employee Well-Being: OHSMS aims to protect and promote the health and well-being of employees. This includes addressing physical and mental health concerns, providing a safe working environment, and offering support for employees facing occupational health challenges.
- Improved Morale and Productivity: A safe and healthy workplace contributes to improved employee morale and job satisfaction. When employees feel confident in their safety, it positively impacts productivity and engagement, creating a more positive organizational culture.
- Reduced Absenteeism and Turnover: A well-implemented OHSMS can lead to reduced absenteeism due to work-related injuries or illnesses. Employees have a greater inclined to stay in an organization that prioritizes their safety and well-being, which can lead to reduced employee turnover.
- Cost Savings: OHSMS can result in cost savings by preventing workplace accidents and illnesses. This includes reduced workers' compensation claims, lower healthcare costs, and decreased expenses related to absenteeism and turnover.

- Continuous Improvement: OHSMS emphasizes a cycle of continuous improvement. Organizations benefit from regular monitoring, appraisal, and evaluation of their OH&S performance to identify areas for enhancement and implement changes to improve overall safety outcomes.
- Enhanced Reputation: An organization's excellent reputation is enhanced by its intense dedication to occupational health and safety. This can be appealing to customers, investors, and potential employees who value responsible and ethical business practices.
- Supplier and Contractor Management: OHSMS can be applied to the management of suppliers and contractors to ensure that they adhere to the same health and safety standards. This helps organizations maintain a consistent safety culture across their entire supply chain.
- Emergency Preparedness: OHSMS addresses emergency preparedness and response, ensuring that organizations have plans and procedures in place to respond effectively to accidents, natural disasters, or other emergencies that could impact the health and safety of employees.
- Employee Involvement and Empowerment: OHSMS encourages employee participation in safety initiatives. Involving employees in the development and implementation of health and safety programs empowers them to take an active role in their own well-being.

17.3 Introduction to ISO 45001 Standard

ISO 45001 is the global OH&S standard, developed to safeguard employees and individuals from work-related injuries and diseases. It was created by an organization of health and safety management experts that investigated several system management methodologies. Other contemporary standards, such as OHSAS 18001, are also taken into account by ISO 45001 [5]. It is primarily aimed at senior management and is intended to assist businesses in providing a healthy and safe working environment for their workers and visitors. This is accomplished by minimizing situations that might cause injury, sickness, or death. ISO 45001 is a watershed moment in occupational health and safety, providing organizations of all sizes with a consistent framework for building more robust and more effective safeguards. It is an important turning point in workplace health and safety [6].

17.3.1 ISO 45001:2018 Standard and Its Key Elements

ISO 45001:2018 is an international standard that specifies requirements for an OHSMS. The standard was developed by the International Organization for Standardization (ISO) and published in March 2018. ISO 45001 can be used by organizations to provide a structure for regulating OH&S hazards and enhancing overall safety management. This structure aims to enhance consistency and compatibility among different management systems, making it easier for organizations to integrate multiple systems. ISO 45001:2018 is intended to assist organizations in creating safer workplaces, preventing work-related accidents and illnesses, and continuously improving their OH&S performance. It also serves as a foundation for organizations to show to stakeholders their support for occupational health and safety, including employees, customers, and regulatory authorities. Organizations seeking certification to ISO 45001:2018 undergo an audit process to assess conformity with the standard's requirements [7]. Key aspects of ISO 45001 are as follows:

1. Scope and Applicability: ISO 45001 applies to organizations of all sizes and industries. It is designed to be adaptable to different types of organizations, including those with complex structures or operating in multiple locations.
2. Context of the Organization: The standard emphasizes the necessity of knowing the organization's internal and external information, including the requirements and demands of those involved. This knowledge guides the creation of the OH&S system of management.
3. Leadership and Commitment: Demonstrate leadership commitment to OH&S by establishing an OH&S policy, defining roles and responsibilities, and ensuring the availability of necessary resources. Top management must exhibit leadership and dedication to the OH&S system of management. Worker participation and consultation are integral to the effective implementation of the system.
4. Policy and Objectives: Develop an OH&S policy aligned with organizational goals. Set measurable objectives and targets for continual improvement.
5. Risk Assessment and Hazard Identification: Conduct systematic risk assessments and identify hazards to determine potential OH&S risks. Implement controls to mitigate risks and ensure the health and safety of workers.

6. Legal Compliance: Create systems for identifying and adhering to relevant laws and regulations for occupational health and safety.
7. Consultation and Participation: Encourage employee involvement and consultation in OH&S matters, including decision-making processes and hazard identification. ISO 45001 is intended to connect seamlessly to additional ISO management system standards. This facilitates the development of integrated management systems.
8. Operational Controls: Implement controls and procedures to manage identified risks, including safe work practices, emergency response, and monitoring of operational activities.
9. Emergency Preparedness and Response: Create strategies and processes for successfully responding to crises, accidents, and events that could have an impact on occupational health and safety.
10. Performance Monitoring and Measurement: Establish processes for monitoring and measuring OH&S performance using key performance indicators (KPIs) and other relevant metrics. ISO 45001 requires organizations to monitor, measure, analyze, and evaluate their OH&S performance. This includes monitoring compliance, incident investigation, and performance indicators.
11. Incident Investigation: Develop a systematic investigation method for accidents, events, and near misses. Determine the fundamental reasons and take appropriate measures to avoid recurrence.
12. Performance Evaluation: Monitor the OHSMS's performance against defined goals and objectives on a regular basis. Conduct management reviews to assess overall effectiveness.
13. Continual Improvement: Encourage a culture of continuous improvement by recognizing areas for development and adopting preventive and remedial measures. Organizations are required to enhance the appropriateness, sufficiency, and effectiveness of their OHS governance structure on a continuous basis. This involves responding to incidents, taking corrective actions, and making enhancements based on performance evaluation.
14. Documentation and Record Keeping: Establish and maintain documentation and records necessary to support the OHSMS and demonstrate compliance with the standard.

The standard emphasizes the importance of maintaining documented information, including policies, procedures, and records, to demonstrate conformity to the requirements of ISO 45001.

15. Worker Participation and Consultation: Encourage active worker participation in OH&S processes and decisions. Provide mechanisms for workers to express concerns and contribute to the improvement of safety.
16. Supplier and Contractor Management: Integrate OH&S considerations into the selection and management of suppliers and contractors, ensuring alignment with the organization's safety standards.

17.3.2 Benefits of ISO 45001 Certification

Achieving ISO 45001 certification involves a rigorous process of implementing and maintaining the OH&S MS, undergoing audits by certified bodies, and demonstrating compliance with the standard's requirements. Certification is a testament to an organization's commitment to creating a safe and healthy workplace for its employees. Few advantages are as follows:

- Improved OH&S Performance: ISO 45001 establishes a method for addressing OH&S hazards, leading to improved safety performance and a safer working environment.
- Legal Compliance: Certification demonstrates a commitment to complying with relevant permissible and governing requirements associated to occupational health and safety.
- Enhanced Reputation: ISO 45001 certification enhances the organization's reputation, signaling to stakeholders, customers, and the public that the organization prioritizes the well-being of its employees.
- Global Recognition: ISO 45001 is an internationally recognized standard, providing a common language and framework for OH&S management. This recognition can be advantageous in global markets.
- Competitive Advantage: Certification can be a differentiator in the marketplace, providing a competitive advantage, especially in industries where clients and customers prioritize safety in their selection criteria.

- Employee Engagement: ISO 45001 emphasizes worker participation and consultation, fostering a culture of safety and engagement among employees.
- Reduced Incidents and Accidents: ISO 45001 can help to reduce the incidence and extent of workplace accidents and incidents by systematically identifying and resolving OH&S concerns.
- Cost Savings: Proactive risk management and a focus on prevention can lead to cost savings by minimizing workplace injuries, reducing absenteeism, and improving overall operational efficiency.
- Continual Improvement: ISO 45001 encourages organizations to routinely review their achievements, set targets, and take steps to improve their OH&S system of management while promoting a culture of continuous improvement.
- Worker Involvement: The standard emphasizes the importance of involving workers in safety processes, contributing to a collaborative approach to identifying and addressing safety concerns.
- Improved Communication: The certification process often involves improving communication about safety policies, procedures, and performance both within the organization and with external stakeholders.

17.3.3 Drawbacks of ISO 45001 Certification

When considering ISO 45001 certification, organizations should carefully weigh the benefits against the potential challenges and ensure that the decision aligns with their overall business objectives and priorities. Additionally, a successful implementation should involve commitment from leadership, engagement of employees, and a focus on continuous improvement rather than just meeting certification requirements. Few disadvantages are as follows:

- Resource Intensive: Achieving and maintaining ISO 45001 certification can be resource-intensive, requiring investments in time, personnel, and financial resources.
- Complex Implementation: Implementing ISO 45001 can be complex, especially for organizations with limited experience in formal management systems. It may require changes to existing processes and the development of new documentation.

- Ongoing Commitment: Maintaining certification requires a continuous commitment to compliance, which may pose challenges for organizations facing competing priorities or resource constraints.
- Resistance to Change: Employees and management may resist changes to established work practices or the introduction of new procedures associated with ISO 45001 implementation.
- Perceived Bureaucracy: Some organizations may perceive the documentation requirements of ISO 45001 as bureaucratic and administrative, leading to concerns about the added workload.
- Initial Certification Costs: The costs associated with initial certification, including the hiring of external auditors and meeting certification requirements, can be significant.
- Misinterpretation of Requirements: Misinterpretation of ISO 45001 requirements can lead to the development of ineffective safety management practices that focus more on meeting the standard's criteria than on improving safety outcomes.
- Limited Flexibility: The standard's requirements may not fully align with certain organizational structures or industries, leading to concerns about the flexibility of the framework.
- Potential for Overemphasis on Certification: There is a risk that organizations may focus more on achieving and maintaining certification than on genuinely improving safety performance and creating a safety culture.
- Not a Guarantee of Safety: ISO 45001 certification is a tool for managing safety but does not guarantee a hazard-free workplace. Organizations must remain vigilant in identifying and addressing emerging risks.

17.3.4 Application of ISO 45001 Standard

The ISO 45001 standard, which focuses on OHSMS, can be applied across various industries and organizations. Its application is designed to help organizations systematically manage OH&S risks, prevent work-related injuries and illnesses, and create safer working conditions. The application of ISO 45001 is versatile, and organizations can tailor its principles to their specific needs and contexts. It provides a systematic and structured approach to managing occupational health and safety, contributing to the overall well-being of employees and the success of the organization [8]. Few applications of the ISO 45001 standard are as follows:

- Risk Management: ISO 45001 offers organizations with a systematic framework for identifying, assessing, and managing OH&S hazards. By understanding and addressing potential hazards, organizations can minimize the likelihood of accidents and incidents.
- Improving OH&S Performance: The standard supports organizations in establishing and implementing processes that lead to continual enhancement in OH&S performance. Regular assessment and review help identify areas for enhancement.
- Employee Involvement: ISO 45001 encourages active involvement of employees in the improvement, implementation, and development of the OHSMS. Engaging employees in safety initiatives fosters a culture of responsibility and shared commitment to safety.
- Emergency Preparedness and Response: The standard addresses emergency preparedness and response, ensuring that organizations have plans in place to effectively respond to emergencies that could impact the health and safety of employees.
- Supplier and Contractor Management: ISO 45001 can be applied to the management of suppliers and contractors to ensure that they adhere to the same OH&S standards. This extends the organization's safety culture throughout its supply chain.
- Enhanced Leadership and Governance: ISO 45001 emphasizes leadership assurance to occupational health and safety. The application of the standard encourages organizations to establish a strong governance structure and leadership commitment to drive a positive safety culture.
- Reducing Workplace Incidents: Through the identification and control of hazards, implementation of preventive measures, and ongoing monitoring, the application of ISO 45001 aims to reduce workplace incidents and improve the overall safety record of an organization.

- Employee Well-being: ISO 45001 contributes to the improvement of employee well-being by addressing both physical and mental health aspects.
- Demonstrating Commitment to Stakeholders: Certification to ISO 45001 provides organizations with a tangible way to demonstrate their assurance to OH&S to stakeholders, including employees, customers, regulatory authorities, and the public.
- Continuous Improvement Culture: The standard encourages a culture of continuous development by demanding organizations to regularly assess and review their OH&S performance and make adjustments to enhance effectiveness.
- Enhancing Organizational Reputation: By implementing ISO 45001 and achieving certification, organizations can enhance their reputation by showcasing an assurance to creating a safe and healthy work environment, which can be a competitive advantage.

17.3.5 How Can the Efficiency of OHSMS Be Increased?

To make OHSMS more efficient, organizations can implement various strategies and practices that enhance the effectiveness of the system and improve overall safety performance. Organizations can enhance the efficiency of their OHSMS, reduce risks, and create a safer and healthier work environment for employees. Regular reviews and adjustments based on performance data and feedback contribute to the ongoing improvement of the OHSMS. Key steps to increase the efficiency of OHSMS:

- Leadership Commitment: Show a clear and ongoing commitment to workplace health and safety as a leader. Leaders must set an example to follow, emphasize the significance of safety, and participate actively in safety efforts.
- Employee Involvement: Engage employees at all levels in the OHSMS. Incorporate them actively in the identification of hazards, assessment of risks, and the creation of safety rules and procedures. Employees' input and participation enhance the effectiveness of the system.
- Regular Training and Education: Employees should be given regular OH&S training and instruction. Ensure that personnel are informed of safety rules and processes, and the importance of adhering to safe work practices.

- Continuous Improvement Culture: Create a culture of constant development. Encourage staff to determine areas for development, report hazards, and actively participate in safety initiatives. Regularly review and update OHSMS procedures based on lessons learned.
- Proactive Hazard Identification: Implement proactive hazard identification processes. Encourage regular workplace inspections, risk assessments, and the reporting of near misses. This helps in identifying potential risks before they lead to incidents.
- Effective Communication: Establish clear and effective communication channels for sharing safety information. Ensure that employees receive timely and relevant information about safety policies, procedures, and any changes in the OHSMS.
- Performance Metrics and Monitoring: Define and monitor key performance metrics for OH&S on a regular basis. Metrics can be used to analyze the efficacy of management policies, event rates, and overall safety effectiveness.
- Use of Technology: Leverage technology to streamline OHSMS processes. Implement safety management software, incident reporting systems, and data analytics tools to enhance data collection, analysis, and reporting.
- Incident Investigation and Learning: Investigate events and near misses thoroughly and promptly. Concentrate on identifying fundamental causes and adopting preventative measures to avoid repetition. Use these investigations as opportunities for organizational learning.
- Legal Compliance: Stay informed about and comply with applicable OH&S regulations and standards. Regularly review and update procedures to ensure alignment with legal requirements.
- Emergency Preparedness: Maintain effective emergency preparedness and response plans. Regular drills should be conducted to make sure that personnel are aware with emergency protocols and are able to react efficiently in the event of an accident or crisis.
- Integration with Other Management Systems: Connect the OHSMS with additional systems of management, such as environmental and quality management. This integration enhances efficiency and promotes a holistic approach to organizational management.

- Employee Wellness Programs: Implement wellness programs that address both physical and mental health. A focus on employee well-being contributes to a healthier workforce and can positively impact safety and productivity.
- Regular Management Review: Assess the system's performance, review data and trends, and make informed decisions on areas for improvement and resource allocation.
- Supplier and Contractor Management: Extend safety considerations to suppliers and contractors. Ensure that they adhere to safety standards and integrate them into the overall OHSMS. Include safety performance criteria in the selection and evaluation of suppliers and contractors.
- Documentation and Record Keeping: Streamline documentation and record-keeping processes. Ensure that relevant information is documented, controlled, and easily accessible. This includes safety policies, procedures, incident reports, and training records.
- Benchmarking: Benchmark OHSMS performance against industry standards and best practices. Identify opportunities for improvement by comparing safety performance with similar organizations.
- Regular Audits and Inspections: Perform frequent institutional inspections and audits to ensure OHSMS compliance. Use audit results to determine areas for improvement and to ensure that defined standards are met on an ongoing basis.
- Recognition and Incentives: Implement recognition and incentive programs that reward positive safety behaviors and achievements. This reinforces a safety-focused culture and motivates employees to actively contribute to the OHSMS.
- Feedback Mechanisms: Implement feedback systems to collect employee feedback on the efficiency of the OHSMS. Encourage open communication and address concerns or suggestions promptly.

17.4 Occupational Safety, Health, and Working Conditions Code

The Occupational Safety, Health, and Working Conditions (OSHWC) Code is a labor law in India that was introduced in 2020 to consolidate and streamline various existing labor laws related to OSHWCs [9]. It is part of the labor law reforms undertaken by the Government of India to simplify and modernize labor legislation. The removal of employment legislation and compliance will help Indian businesses. Organizations must ensure that they are in conformity with the new legislation. The codes' implementation is continuously monitored, and any concerns or obstacles that occur are dealt with in a rapid and efficient way.

17.4.1 Rights of Employees Under the OSHWC Code

In India, the OSHWC Code intends to integrate and simplify current labor legislation concerning OSHWCs. The OSH Code provides various rights and protections for employees in the context of occupational safety and health. It is crucial for employees to be aware of their rights under the OSH Code and for employers to comply with its provisions to ensure a safe and healthy work environment [10]. Additionally, changes to labor laws and regulations may occur, so employees should stay informed about any updates or amendments to the OSH Code. Some key rights of employees under the OSH Code are as follows:

- Safe and Healthy Working Conditions: Employees have the opportunity to work in a safe and healthy workplace. The employer is responsible for providing and maintaining facilities, equipment, and conditions that ensure the health and safety of employees.
- Right to Know: Employees have the right to be informed about the potential hazards associated with their work. Employers are required to communicate information about workplace risks, safety procedures, and measures to protect against occupational hazards.
- Right to Training: Employees have the right to receive training on occupational safety and health practices. Employers are responsible for providing necessary training to ensure that employees are aware of safety protocols and can perform their jobs safely.
- Right to Personal Protective Equipment (PPE): Employees have the right to be provided with suitable personal protective equipment where necessary. Employers must supply and maintain PPE to protect employees from workplace hazards.
- Right to Medical Examination: The OSH Code provides provisions for employees to undergo periodic health check-ups.

Employers may arrange for medical examinations to monitor the health of employees exposed to specific occupational risks.

- Right to Rest and Overwork Limitations: Employees have the right to rest periods and limitations on working hours. The OSH Code establishes standards for daily and weekly working hours, rest intervals, and overtime, with the aim of preventing fatigue and ensuring employee well-being.
- Right to Leave and Holidays: The code outlines provisions related to various types of leave, including annual leave with wages, sick leave, and maternity leave. Employees have the right to take leave as per the stipulated regulations.
- Right to Compensation: In case of work-related injuries, illnesses, or accidents resulting in disability or death, employees have the right to compensation. Employers are obligated to provide compensation as per the provisions of the code and relevant laws.
- Right to Representation: The OSH Code allows for the establishment of safety committees and the appointment of safety officers. Employees have the right to be represented on safety committees, contributing to decision-making processes related to occupational safety and health.

17.4.2 Rights of Employers Under the OSHWC Code

The OSHWC Code in India, like many labor laws, outlines both the rights and responsibilities of employers. It is important to note that, while employers have rights, these rights are balanced by corresponding responsibilities. The OSH Code and other labor laws also outline the obligations of employers to provide a safe and healthy working environment, adhere to labor standards, and respect the rights of workers. Additionally, labor laws are subject to change, and specific rights and obligations may vary based on jurisdiction and industry. Employers are advised to stay informed about applicable labor laws and seek legal counsel to ensure compliance with the latest regulations. Some of the key rights of employers under the OSH Code are as follows:

- Management and Control: Employers have the right to manage and control the workplace, including decisions related to production, operations, and the overall functioning of the organization.

- Setting Policies and Procedures: Employers have the right to establish and enforce workplace policies and procedures, including those related to occupational safety and health. This involves creating a framework for safety practices and ensuring compliance with regulatory requirements.
- Employment Contracts: Employers have the right to enter into employment contracts with workers. These contracts typically outline the terms and conditions of employment, including roles, responsibilities, working hours, and remuneration.
- Discipline and Termination: Employers have the right to discipline employees for violations of workplace policies, code of conduct, or poor performance. They also have the right to terminate employment, subject to the provisions of applicable labor laws.
- Hiring and Recruitment: Employers have the right to make decisions regarding the hiring and recruitment of employees, based on qualifications, skills, and other relevant criteria.
- Health and Safety Measures: Employers have the right to implement health and safety measures in the workplace to ensure the well-being of employees. This includes conducting risk assessments, providing safety training, and implementing safety protocols.
- Information and Training: Employers have the right to provide information and training to employees, particularly regarding occupational safety and health practices. This is essential for creating awareness and ensuring compliance with safety regulations.
- Collective Bargaining: Employers have the right to engage in collective bargaining with employee representatives or trade unions to negotiate terms and conditions of employment, which includes discussions related to working hours, wages, and other matters related to employment.
- Flexibility in Work Arrangements: Depending on the nature of the work and organizational requirements, employers have the right to implement flexible work arrangements, such as shift schedules, remote work, or other arrangements that meet operational needs.
- Implementation of OSH Code Compliance: Employers have the responsibility and corresponding right to implement measures that ensure compliance with the provisions of

the OSH Code. This includes maintaining health and safety records, conducting inspections, and addressing any identified hazards.

17.4.3 Health, Safety, and Working Conditions of Employees Under the OSH Code

The OSH Code aims to consolidate and streamline existing labor laws related to OSHWCs. Key provisions related to the health, safety, and working conditions of employees under the OSH Code are as follows:

- Scope and Applicability: The OSH Code applies to establishments with 10 or more employees, like factories, mines, plantations, construction sites, and other establishments specified by the appropriate government.
- Working Hours: The code specifies standard working hours, intervals for rest, and conditions for overtime work. It addresses daily and weekly working hours, spread-over, and rest intervals to ensure the well-being of employees.
- Health and Safety: The OSH Code emphasizes the importance of maintaining a safe and healthy working environment. It includes provisions related to ventilation, temperature, lighting, sanitation, and other factors which affect workers safety and health.
- Welfare Measures: Employers are obligated to provide various welfare measures, including first aid facilities, canteens, restrooms, and facilities for storing and drying clothing. These measures contribute to the overall well-being of employees.
- Annual Health Check-ups: Regular health check-ups by the establishments help in early detection of health issues and contribute to preventive healthcare.
- Occupational Safety and Health Committees: The code envisages the establishment of occupational safety and health committees to promote a participative approach to ensuring a safe working environment. These committees involve both employers and employees in matters related to health and safety.
- Leave and Holidays: The OSH Code specifies provisions related to leave entitlements, including annual leave with

wages, maternity leave, and other types of leave. These provisions contribute to the overall well-being and work-life balance of employees.

- Social Security: The code includes provisions related to social security for workers, covering aspects such as provident fund, employee state insurance, and gratuity. Social security measures aim to provide financial protection to employees and their families.
- Contract Labor: Provisions related to contract labor are included to regulate the engagement of contract workers and ensure their welfare and safety. This includes measures to prevent the exploitation of contract labor.
- Penalties: The OSH Code specifies penalties for failing to comply with its rules and regulations. According to the type and degree of the infraction, penalties can involve fines and imprisonment. Strict enforcement aims to ensure adherence to health and safety standards.

17.4.4 Applications of OSHWC Code

The OSHWC Code has wide-ranging applications across industries and workplaces in the country. The OSH Code is designed to bring clarity, uniformity, and simplicity to labor laws in India, ensuring that workers across different sectors receive fair treatment and protection. Its application is vital for organizations to adhere to legal requirements, foster a safe working environment, and contribute to the overall well-being of the workforce. Some key applications of the OSHWC Code are as follows:

- Safety and Health Standards: The OSH Code establishes safety and health standards for different types of establishments, including factories, mines, construction sites, and more. It sets out the requirements that organizations must follow to ensure a safe and healthy working environment for employees.
- Working Hours and Conditions: The code addresses working hours, rest intervals, and other conditions of work. It establishes standards for daily and weekly working hours, breaks, and overtime, aiming to safeguard the well-being of workers and prevent overwork.
- Welfare Measures: It outlines welfare measures that employers are required to provide for employees, including facilities

like canteens, restrooms, first aid, and more. These measures are designed to enhance the overall well-being of workers.

- Leave Entitlements: The code specifies provisions related to various types of leave, such as annual leave with wages, maternity leave, and other leave entitlements. These provisions are designed to support employees during different life events.
- Contract Labor: The code includes provisions related to the engagement of contract labor. It aims to regulate the working conditions and rights of contract workers, ensuring fair treatment and compliance with labor laws.
- Social Security: The code addresses social security aspects for workers, including provisions related to provident fund, employee state insurance, and gratuity. These measures contribute to the financial well-being of workers.
- Annual Health Check-ups: Employers are required to provide annual health check-ups for certain categories of workers. This provision supports preventive healthcare and early detection of health issues among employees.
- Electronic Record-Keeping: The code introduces electronic record-keeping requirements for maintaining records related to wages, work hours, and other relevant information. This move toward digitization aims to enhance transparency and accessibility of records.
- Single License for Multiple Establishments: The code allows for a single license to be issued for multiple establishments owned by the same employer. This streamlines administrative processes for organizations operating in multiple locations.
- Inspection and Compliance: The code empowers inspectors to conduct inspections to ensure compliance with its provisions. This regulatory oversight helps in enforcing standards and holding organizations accountable for the well-being of their workers.

References

1. Makin, A.M. and Winder, C., A new conceptual framework to improve the application of occupational health and safety management systems. *Saf. Sci.*, 46, 6, 935–948, 2008.

2. Çalış, S. and Büyükakıncı, B.Y., Occupational health and safety management systems applications and a system planning model. *Procedia Comput. Sci.*, 158, 1058–1066, 2019.
3. Mohammadfam, I., Kamalinia, M., Momeni, M., Golmohammadi, R., Hamidi, Y., Soltanian, A., Evaluation of the quality of occupational health and safety management systems based on key performance indicators in certified organizations. *Saf. Health Work*, 8, 2, 156–161, 2017.
4. Robson, L.S., Clarke, J.A., Cullen, K., Bielecky, A., Severin, C., Bigelow, P.L., Irvin, E., Culyer, A., Mahood, Q., The effectiveness of occupational health and safety management system interventions: a systematic review. *Saf. Sci.*, 45, 3, 329–353, 2007.
5. Darabont, D.C., Antonov, A.E., Bejinariu, C., Key elements on implementing an occupational health and safety management system using ISO 45001 standard, in: *MATEC web of conferences*, vol. 121, EDP Sciences, 2017.
6. Karanikas, N., Weber, D., Bruschi, K., Brown, S., Identification of systems thinking aspects in ISO 45001: 2018 on occupational health & safety management. *Saf. Sci.*, 148, 105671, 2022.
7. Darabont, D.C., Bejinariu, C., Ionita, I., Bernevig-Sava, M.A., Baciu, C., Baciu, E.R., Considerations On Improving Occupational Health And Safety Performance In Companies Using ISO 45001 Standard. *Environ. Eng. Manage. J. (EEMJ)*, 17, 11, 2711–2717, 2018.
8. Uzun, M., Gurcanli, G.E., Bilir, S., Change in occupational health and safety management system: ISO 45001: 2018, in: *5th International Project Management and Construction Conference (IPCMC 2018)*, Cyprus International University, North Cyprus, 2018.
9. Jain, M. and Jain, M., Critical Analysis of Occupational Safety, Health Working Conditions Code Bill 2020. *Int'l. JL Mgmt. Hum.*, 4, 2, 2136, 2021.
10. Pujari, A., A Critical Analysis" the Occupational Safety, Health and Working Conditions Code, 2020". *Indian JL Legal Rsch.*, 3, 1, 2021.

Index

Also of Interest

Check out these related titles from Scrivener Publishing

Disaster Management and Environmental Sustainability, Edited by Sanjay Kumar, Suraj Kumar Singh, Shruti Kanga, Gowhar Meraj, Majid Farooq, and M.S. Nathawat, ISBN: 9781394166350. Learn how environmental degradation is driving disaster risk and discover strategies to minimize it through environment-sensitive development in this insightful book.

LAND AND ENVIRONMENTAL MANAGEMENT THROUGH FORESTRY, Edited by Abhishek Raj, Manoj Kumar Jhariya, Arnab Banerjee, Sharad Nema, and Kiran Bargali, ISBN 9781119910404. Written and edited by a group of experts in the field, this groundbreaking reference work sets the standard for engineers, students, and professionals working in forestry, agriculture, ecology and environmental science, offering the scientific community a way towards combating climate change.

GREEN CHEMISTRY FOR ENVIRONMENTAL REMEDIATION, Edited by Rashmi Sanghi and Vandana Singh, ISBN: 9780470943083. Presents high quality research papers as well as in depth review articles on the new emerging green face of multidimensional environmental chemistry.

BIOREMEDIATION AND SUSTAINABILITY: Research and Applications, Edited by Romeela Mohee and Ackmez Mudhoo, ISBN: 9781118062845. Bioremediation and Sustainability is an up-to-date and comprehensive treatment of research and applications for some of the most important low-cost, "green," emerging technologies in chemical and environmental engineering.

EMERGENCY RESPONSE MANAGEMENT OF OFFSHORE OIL SPILLS: Guidelines for Emergency Responders, By Nicholas P. Cheremisinoff, PhD, and Anton Davletshin, ISBN: 9780470927120. The first book to examine the Deepwater Horizon disaster and offer processes for safety and environmental protection.

COMPLEX BIOLOGICAL SYSTEMS: Adaptation and Tolerance to Extreme Environments, By Irina R. Fomina, Karl Y. Biel, and Vladislav G. Soukhovolsky, ISBN: 9781119510406. Written and edited by some of the most well-respected authors in the area of the adaptation of plants and animals to climate change, this groundbreaking new work is an extremely important scientific contribution to the study of global warming.

GLOBAL CLIMATE CHANGE DEMYSTIFIED, By James G. Speight, ISBN: 9781119653851. Tackling one of the most controversial subjects of our time, one of the world's foremost environmental and petroleum engineers explores the potential causes and ramifications of global climate change.

DEATH OF THE CHESAPEAKE: A History of the Military Role in Polluting the Bay, By Richard D. Albright, ISBN: 9781118686270. This unique book focuses attention on the failure of current efforts to cleanup the Chesapeake Bay and suggests an approach often used in cleaning up environmentally damaged sites.

Printed and bound by CPI Group (UK) Ltd, Croydon, CR0 4YY

15/09/2026

14962409-0002